Pollution, Congestion, and Nuisance

Pollution, Congestion, and Nuisance

The Economics of Nonmarket
Interdependence

Daniel T. Dick
University of Santa Clara

Lexington Books
D.C. Heath and Company
Lexington, Massachusetts
Toronto   London

Library of Congress Cataloging in Publication Data

Dick, Daniel T.
    Pollution, congestion, and nuisance; the economics of nonmarket
interdependence.

    Bibliography: p.
    1. Economics. 2. Mathematical optimization. 3. Equilibrium (Eco-
nomics) I. Title.
HB199.D495   330   73-15461
ISBN 0-669-91165-8

Published simultaneously in Canada.

Printed in the United States of America.

International Standard Book Number: 0-669-91165-8

Library of Congress Catalog Card Number: 73-15461

For my parents,
Jacob and Leona Dick

Table of Contents

Preface

The purpose of this book is to sort out various ways of modeling externality relationships with emphasis on the applicability of various models, to put in perspective some of the controversies that have arisen in the literature, and to provide a reference for issues surrounding the efficiency property of policies designed to deal with externalities. In pursuing this purpose I have attempted to stick closely to original source arguments and to document them liberally but honestly. Yet, I have added enough "textbook-type" explication to allow the work to stand on its own as a self-contained unit.

One underlying theme of the book is that there is a great deal microeconomics has to say about the externality problem. As distinct from chemical, physical, or biological approaches to environmental problems, the economic approach treats the problem from the point of view of allocation and resource use and the incentives guiding allocative decisions. This leads to the second underlying theme, and that is that the spirit if not the letter of the Pigovian analysis is correct. Commoner's observation that modern technology has been counter-ecological is right, but it has been counter-ecological for a reason. And that is the price vector faced by those making decisions about technology is not truly reflective of nonmarket scarcities. A third thread tying the work together is really an epistemological one, and that is that our knowledge of complex social phenomena is limited by our knowledge of models about these phenomena.[1] A striking example comes from the history of macroeconomic theory and policy prescriptions for dealing with unemployment.

As with any subject area, there is a large volume of literature which blends imperceptibly into the material that is treated. The choice of what to include or exclude is based largely on my own judgment as to the immediate relevance of material to the core of the subject area, keeping in mind the space constraint. However, to partially compensate, I have included a bibliography which is rather more extensive than the works directly referenced in the text and which will probably be of interest to anyone whose interest has led him to this work.

Two details need to be clarified about word usage in the text. Nuisance has a specialized use which is defined in the text and which differs from legal and everyday usages. The ordinate of most of the graphs is labeled value. This is not price times quantity, but rather the units in which value is measured, e.g., dollars per unit, or francs per unit.

Finally, it is a pleasure to acknowledge people without whose help this work would have never been undertaken or completed. Mario Belotti, Chairman, Department of Economics, University of Santa Clara, deserves my special thanks for permitting me the freedom and flexibility in both course scheduling and content that is essential if research is to be undertaken. I owe a great debt to a group of men in the Economics Department at the University of California,

Riverside: Ralph D'Arge, Paul Downing, and Kay Hunt, who had a profound influence on the direction of my interests during my brief contact with them in 1969-70. I would like to thank my colleagues, Henry Demmert, Department of Economics, for providing valuable discussions on various aspects of the subject, and Stein Weissenberger, Department of Mechanical Engineering, currently on leave at NASA-Ames Research Center, Mountain View, California, for providing me with an appreciation of what little control theory I do know, and for in effect coauthoring the material from which the second section of Chapter 7 is taken. Esmail Amid-Hozour, Finance Ministry, Teheran, Iran, was also helpful in the formulation of the pollution control problem. I am indebted to three very capable research assistants, Richard Burdick, Robert Finnochio, and Thomas Quinlan, formerly of the University of Santa Clara. Christine Woodward deserves my special thanks for typing the manuscript accurately and efficiently and for meeting my deadlines cheerfully and promptly. Finally, I owe my wife, Cathryn, a deep debt of gratitude, for understanding the demands of authorship and for the help she has been during this undertaking.

1 Introduction

That there is abuse of our natural environment is a daily fact of life for much of the urban population in industrialized countries. The December 1952 smog to which is attributed more than four thousand excess deaths in one week in London is a dramatic and spectacular reminder of the day-in and day-out costs that air pollution imposes in a less spectacular way.[1] It has been estimated that a 50 percent reduction in air pollution throughout this country would increase life expectancy of a child born in 1970 by three to five years. Alternatively, the saving in missed working days and decreased direct expenditure on health care flowing from a similar reduction in air pollution is estimated to be $2 billion annually.[2] The total cost of air pollution due to deterioration in human health, damage to plant life and material, and reduced property values is estimated to be about $16 billion annually in the United States.[3] The fact is that pollution is not only a problem in the capitalist countries of the industrialized west, but also in a country like the USSR in which the state owns and controls the means of production. Goldman chronicles the environmental disruption in the Soviet Union replete with stories of burning rivers, oil slicks, and fish kills.[4] The relevant question with regard to using environmental resources in both the USSR and the United States is not who should own the means of production, but rather how can resources be used to minimize the true cost of pollution, which is the damage cost plus the abatement cost, and secondly how can economic decisionmakers be induced to achieve this resource use.

Choice of Criterion

This leads immediately to the criterion with respect to which optima are defined and policies are evaluated, and that is the Paretian criterion of maximum social product. Despite the widespread use of this criterion in the writings of applied microeconomists, it is good to recognize that the choice of this criterion represents a value judgment. There are other criteria that could be used, such as improvement in the income distribution,[5] improvement of the balance of payments, or some weighted average of these. Therefore, it is good to note that in the analysis which follows, the selection of the Paretian criterion is a value judgment (or the selection of a norm), with respect to which valuations of alternative actions are made. Words like "proper allocation," or "suboptimal allocation" only have meaning with respect to a norm, the selection of which represents a value judgment.[6] Therefore in a very real sense the analysis is value-laden and not value-free.

Outline of the Book

Chapters 1 and 2 constitute a preface to the core of the material which follows. Chapter 2 reviews the optimal property of markets when decisionmakers have no direct, nonmarket interdependence, and establishes a notation and the use of the Paretian criterion. Chapters 3-6 constitute a part of the book dealing with static, partial equilibrium models. Chapter 3 deals with the nuisance model of externalities in many of its ramifications, while Chapter 4 treats the pollution model and its allocative ramifications. Chapters 5 and 6 deal with the evaluation of policy alternatives in, respectively, the nuisance and pollution case, and both voluntary and involuntary actions are explored in each case. Chapter 7 is written in two parts, each of which represents some main features of two larger subject areas. One is on the general equilibrium approach with emphasis on the Russell-Spofford model, and the other is on continuously optimal dynamic models. Chapter 8 brings together conclusions that can be drawn and presents them in succinct form.

2

The Theory of Market Behavior

Introduction

This chapter will examine the allocative problem when all decision-making units are independent of each other and are linked only by the market in a competitive setting. We will show that private optimizers acting selfishly in this setting not only solve the private problem of making the optimizer as well off as possible, but also solve the social problem of making the society as well off as it can be. This is the efficiency property about market behavior that had such great intellectual appeal to Adam Smith and his followers. If each decisionmaker optimizes with respect to his own self-interest, then the society will be led as if by an invisible hand to its social optimum. It is this proposition which will be referred to as the Neoclassical Welfare Theorem that is set forth in this chapter.

More precisely, the Neoclassical Welfare Theorem may be stated as follows:

> *THEOREM. An equilibrium of a competitive pricing mechanism results in a Pareto optimal allocation of resources; any Pareto optimal allocation of resources can be achieved by the equilibration of a competitive pricing system.* [1]

The first section of the chapter presents the problem of attaining a social optimum as one of solving a programming problem, and its solution is shown to be a Pareto optimum. Next, the first part of the theorem is shown to be true for consumption and production economy with given resources, and, finally, motivation for the last part of the theorem is developed.

The Social Optimum as a Programming Problem

In this section, we will set up and solve the problem of allocating given outputs among consumers in a way that will maximize social welfare and the problem of allocating given inputs among production processes in a way that will maximize the value of output. With these twin problems solved, a simple, fixed-resource production-consumption economy will achieve its social optimum.

3

Social Optimum in Consumption

Imagine a benevolent planning agency which has as its task the distribution of fixed amounts of various goods among several consumers. The object of this agency is to make its final distribution so that the total happiness of the society is as great as it can be. But how will the happiness be measured? Each consumer is supposed to possess a utility function which shows the utility to him of various combinations of output which he might consume. For example, consumer A's utility function is[a]

$$U^a = U^a(x_1^a, x_2^a) \qquad (2.1)$$

where x_i^a is the quantity of the ith output consumed by consumer A. We will assume that the arguments, x_i^a, are perfectly divisible; that more is always preferred to less as registered by the index U^a, i.e., the consumer is never satiated in x_1 and x_2, or U^a never reaches a top (bliss point); and also that U^a is strictly concave. These last two assumptions may be most simply stated by placing conditions on the derivatives of U.

$$\partial U^a/\partial x_i^a > 0, \ \partial^2 U^a/(\partial x_i^a)^2 < 0, \text{ all } x_i^a \geqslant 0 \qquad (2.2)$$

We also assume what Arrow termed as being probably "... the most critical [assumption] ..."[2] and that is Arrow's

ASSUMPTION 2. The desirability of a distribution [x_1^a, x_2^a] to individual [A] is solely dictated by the desirability to him [my emphasis] of the commodity bundle [x_1^a, x_2^a].[3]

This establishes the independence of consumers; they are only linked by the market and have no nonmarket, or external, link. Indeed, it is the examination of optimality when this assumption is violated, corresponding to cases of pollution, nuisance, and congestion, that is of interest in the remainder of this work.

If we also suppose consumer B to possess a utility function with the same properties as A's, though not necessarily the same function, then we are in a position to formulate the problem of allocation in consumption. Let the planning agency have

[a]For expository purposes, a world with only two commodities and two consumers will be treated. Generalizations to n-dimensional commodity spaces, n consumers, are immediate. Even infinite-dimensional commodity spaces have been successfully treated in the literature. See G. Debreu, "Valuation Equilibrium and Pareto Optimum," *Proceedings of the National Academy of Sciences of the U.S.A.* 40 (1954): 588-92.

$$x_1 = \bar{x}_1, \; x_2 = \bar{x}_2 \tag{2.3}$$

available to it to distribute to A and B. Its object is to make A and B experience as much utility as possible from the final distribution, or mathematically, to

$$\text{maximize: } [U^a, U^b]. \tag{2.4}$$

The maximization must be performed so that the sum of what is allocated to A and B of each good does not exceed the total amount available, or

$$x_1^a + x_1^b \leqslant \bar{x}_1 \tag{2.5}$$

$$x_2^a + x_2^b \leqslant \bar{x}_2.$$

Finally, of course, there is the nonnegativity requirement

$$x_i^a, x_i^b \geqslant 0 \qquad\qquad i = 1,2. \tag{2.6}$$

The problem of the planning agency in a formal sense is to maximize (2.4) by choosing a nonnegative allocation of goods among the consumers such that no more is distributed than there exists to distribute. But, it is known that a vector maximization problem of this type (maximizing the vector $[U^a, U^b]$) may be reduced to a scalar maximization problem by forming a new maximand which is the sum of positively-weighted elements of the vector.[4] Specifically, the problem of the planning agency becomes to

$$\text{maximize } W(x_1^a, x_1^b, x_2^a, x_2^b) = \alpha^a U^a(x_1^a, x_2^a) + \tag{2.7}$$

$$\alpha^b U^b(x_1^b, x_2^b)$$

$$\text{subject to: } x_1^a + x_1^b \leqslant \bar{x}$$

$$x_2^a + x_2^b \leqslant \bar{x}_2$$

$$x_i^a, x_i^b \geqslant 0 \qquad\qquad i = 1,2,$$

where $\alpha^j > 0, j = a,b,$ and $\sum_{j=a}^{b} \alpha^j = 1$. The α^j's assign weights to the individuals in the welfare function of the society (agency), and Negishi has interpreted them as the reciprocal of the jth consumer's marginal utility of income.[5] Clearly, the weights reflect an ethical judgment about the "deservingness" of the various individuals, and a choice of α's is an arbitrary, ethical one. For example, an egalitarian agency would assign weights of $\alpha^j = 1/2$ to each individual, A and B.

The problem, as stated in (2.7) is a concave programming problem, and as such, the planning agency can rely on the work of Kuhn and Tucker to help it solve the allocative problem of this small, model society. The Kuhn-Tucker Theorem says that the necessary and sufficient condition that X^O [b] be a solution to the problem (2.7) is that there exist some $\Gamma^O \geqslant 0$ [c] such that, for the function

$$L(X, \Gamma) = W(X) + \gamma_1 [x_1 - x_1^a - x_1^b] + \gamma_2 [x_2 - x_2^a - x_2^b] \qquad (2.8)$$

it is true that

$$L(X, \Gamma^O) \leqslant L(X^O, \Gamma^O) \leqslant L(X^O, \Gamma).\text{[6]} \qquad (2.9)$$

Expression (2.9) says that L is at a minimum with respect to Γ and at a maximum with respect to X.

The function L, or so-called Legrangian function, may be interpreted as a payoff function which measures the welfare to society, (W), of a given allocation, (X), plus the value of undistributed resources of that allocation ($\sum_i \gamma_i [x_i - x_i^a - x_i^b]$). Naturally, it is desired to make W as big as possible with respect to X while making the value of the undistributed resources as small as possible. This accounts for the saddle point nature of a solution (X^O, Γ^O) as given by (2.9). In the X "direction" a solution X^O to (2.7) makes L as big as possible; in addition, it is both necessary and sufficient that for X^O to solve (2.7) that a $\Gamma^O \geqslant 0$ exist that makes L as small as possible in the Γ "direction." To complete the interpretation of L as a payoff function in this allocative "game," Γ must be interpreted as a vector of prices corresponding to the goods to be allocated. The solution prices, Γ^O, measure the pure scarcity value of the good in terms of welfare. That is, γ_i^O is the value an extra unit of x_i would afford in creating utility if x_i is allocated between A and B optimally. If some allocation X^O is optimal in which there is x_i left over, then $\gamma_i^O = 0$ because the value in utility terms of another unit of x_i when there is already some slack is zero. Were it not so, there would not be any x_i left over. On the other hand, if an allocation X^O is optimal in which no x_i is left over, then $\gamma_i^O > 0$, and it measures the incremental welfare of an incremental unit of x_i. In either case $\gamma_i [x_i - x_i^a - x_i^b] = 0$, because either $\gamma_i = 0$ or the bracketed term is zero.

Having made the interpretation of L as a payoff function in the allocative "game," and Γ^O as a vector of shadow prices, or pure scarcity prices, of the respective goods available in fixed supply, we can proceed solving the nonlinear program. Actually, we will not so much solve the problem of (2.7) to obtain (X^O, Γ^O), as we will derive conditions which are necessary for (X^O, Γ^O) to make (2.9) true. Moreover, the conditions to be derived are both necessary and

[b]$X^O = [x_1^{ao}, x_2^{ao}, x_1^{bo}, x_2^{bo}]$.

[c]$\Gamma^O = (\gamma_1^o, \gamma_2^o)$.

sufficient when taken together with the assumptions about U^a and U^b that are made.

To derive the conditions, we note that a function achieves its maximum with respect to a set of variables when the first order partial derivates of that function with respect to those variables vanish. Moreover, a function also achieves its minimum with respect to a set of variables when the first order partials with respect to those variables vanish. Therefore, we need only take first partials of (2.8) with respect to all arguments and choose X^O and Γ^O for which these partials vanish.

The first partials of L are as follows:

$$\frac{\partial L}{\partial x_i^a} = \alpha^a \frac{\partial U^a}{\partial x_i^a} - \gamma_i = 0 \qquad\qquad x_i^a = x_i^{ao} \qquad\qquad\qquad (2.10)$$

$$< 0 \qquad\qquad x_i^a = 0, \qquad i = 1,2,$$

$$\frac{\partial L}{\partial x_i^b} = \alpha^b \frac{\partial U^b}{\partial x_i^b} - \gamma_i = 0 \qquad\qquad x_i^b = x_i^{bo}$$

$$< 0 \qquad\qquad x_i^b = 0, \qquad i = 1,2.$$

These six equations can, in principle, be solved for the six variables $X^O = [\, x_1^{ao}, x_2^{ao}, x_1^{bo}, x_2^{bo}\,]$ and $\Gamma^O = (\, \gamma_1^O, \gamma_2^O\,)$. The inequality signs in these conditions add a degree of realism in the solution in that they permit corner optima to exist. That is, each consumer need not receive some of each good from the planning agency in the final allocation. When the model of consumer allocation is extended to many goods and many consumers, it is clear that this property adds realism.

Let us examine the nature of the inequalities a little closer. From the assumptions about the utility indicator, we know that the term $\alpha^a (\, \partial U^a/\partial x_i^a)$ must be positive, but downward sloping throughout the range of $x_i^a \geqslant 0$. This term is A's marginal evaluation of x_i, and its downward slope reflects the diminishing marginal utility that is afforded to A by x_i. If the first equation of (2.10) is met with equality, then the interpretation is that A be provided with x_i to the point where his evaluation of the last unit is just equal to the shadow price of x_i for optimality. Were A provided with one more unit of x_i, his evaluation of that unit is t, whereas that unit has a scarcity value in providing utility elsewhere in the economy of $s = \gamma_i^O$. Therefore, that last unit should be allocated somewhere else in the system, and not to A. The optimal amount of x_i which A should get in order for total welfare to be as large as possible is $x_i^a = x_i^{ao}$.

If, on the other hand, the scarcity value of x_i is everywhere greater than A's marginal evaluation of it, then A should receive none of it. All the x_i can add

8

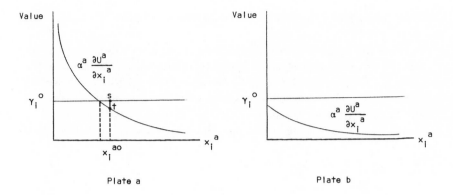

Plate a   Plate b

Figure 2-1.

more to social welfare somewhere else in the economy rather than with A. So, $x_i^a = 0$ when γ_i is greater than A's marginal evaluation of x_i, or when the first condition of (2.10) is met with inequality.

The Arrow paper extends the Neoclassical Welfare Theorem to the case where, "... social optima are in the nature of corner optima ..."[7] This means that the degree of realism added by formulating the allocative problem as a programming problem, with the possibility of allocating none of some goods to some consumers, does not affect the essence of the Welfare Theorem.

The information contained in (2.10) is often presented more compactly as ratios. Assume the first four conditions of (2.10) are met with equality and divide the first by the second equation to get

$$\frac{\alpha^a(\partial U^a/\partial x_1^a)}{\alpha^a(\partial U^a/\partial x_2^a)} = \frac{\gamma_1}{\gamma_2} \tag{2.11}$$

or, cancelling α^a and defining the ratio of A's marginal utilities as his marginal rate of substitution,

$$\text{MRS}_{21}^a = \frac{\gamma_1}{\gamma_2} \tag{2.12}$$

Dividing the third by the fourth equation, we get

$$\frac{\alpha^b(\partial U^b/\partial x_1^b)}{\alpha^b(\partial U^b/\partial x_2^b)} = \text{MRS}_{21}^b = \frac{\gamma_1}{\gamma_2} \tag{2.13}$$

Substituting the solution values of Γ in for γ_1 and γ_2 we get from (2.12) and (2.13)

$$\text{MRS}_{21}^{a} = \text{MRS}_{21}^{b} = \gamma_1^o / \gamma_2^o. \tag{2.14}$$

If X^O is an optimal distribution of output, the marginal rates of substitution between goods must be the same for each individual consuming non-zero amounts of the respective goods. Moreover, the rate of substitution must equal the ratio of the output prices (shadow prices). If the planning agency solves the allocative problem optimally, a solution (X^O, γ^O) will result which makes (2.10), and equivalently (2.14), true.

Social Optimum in Production

Imagine the same benevolent planning agency having to allocate fixed resources to alternative production processes. The object in this case is to make the value of output as big as possible. For purposes of exposition, assume the agency has a fixed amount, $\overline{b}$, of only one primary resource, and that there are two production activities. The output is to be evaluated at prices p_1 and p_2 which gives

$$V = p_1 x_1 + p_2 x_2 \tag{2.15}$$

as the objective to be maximized. Production takes place according to input requirement functions

$$b_1 = f^1 (x_1), \ b_2 = f^2 (x_2) \tag{2.16}$$

where positive amounts of the resource are required for positive output. Moreover, increasing increments of the resource are required for equal increments of output, or

$$\partial f^i / \partial x_i > 0, \ \partial^2 f^i / \partial x_i^2 > 0, \quad i = 1,2, \quad \text{all } x_i \geqslant 0. \tag{2.17}$$

The object is to allocate b between the alternative resource-using production activities in such a way that the value of output is as big as possible, but subject to the constraint that no more of the resource is used than is at the agency's disposal. Formally, the problem is to

$$\text{maximize:} \ V = p_1 x_1 + p_2 x_2 \tag{2.18}$$

$$\text{subject to:} \ f^1 (x_1) + f^2 (x_2) \leqslant \overline{b}$$

$$x_1, x_2 \geqslant 0.$$

Again, the production equivalent of Arrow's Assumption 2 must hold: The amount of input required for production of output of a given process is solely dictated by that output level and not by the level of output of any other process. In other words, there exists no externality.

The problem of (2.18) is a nonlinear programming problem, and therefore we may treat it as we did the consumption problem: Form a payoff function, or Legrangian, L, which gives us the value of output plus the value of unused resources.

$$L(x_1, x_2, \lambda) = p_1 x_1 + p_2 x_2 + \lambda [b - f^1(x_1) - f^2(x_2)]. \quad (2.19)$$

The λ is an unknown multiplier which will be determined as part of a solution. It interprets as the shadow price of the resource, b and as such gives the amounts by which an increment of b would increase the value of output. The Kuhn-Tucker Theorem says that for X^O to be a solution to (2.18) it is necessary and sufficient that there exist a $\lambda^O \geqslant 0$ which makes it true of (2.19) that

$$L(X, \lambda^O) \leqslant L(X^O, \lambda^O) \leqslant L(X^O, \lambda). \quad (2.20)$$

That is, the value of output is maximized (L is maximized in the X direction), while the value of unused resources is minimized (L is minimized in the λ direction).

The necessary, and together with (2.17) also sufficient, condition under which this occurs is that

$$\frac{\partial L}{\partial x_i} = p_i - \lambda(df^i/dx_i) = 0 \qquad x_i = x_i^O \qquad (2.21)$$

$$< 0 \qquad x_i = 0, \qquad i = 1,2,$$

$$\frac{\partial L}{\partial \lambda} = b - f^1 - f^2 \quad = 0 \qquad \lambda = \lambda^O$$

$$> 0 \qquad \lambda = 0.$$

These three equations can, in principle, be solved for $X^O = [x_1^O, x_2^O]$ and λ^O which is the solution to the problem (2.18). As in the consumption problem, the inequalities express the possibility of a corner solution, that is, the planning agency has the option of not allocating the primary resource to one or the other of the production processes.

The interpretation of these conditions is as follows. First note that λ is the shadow price of the resource and that df^i/dx_i is the marginal input requirement. That means that the term $\lambda(df^i/dx_i)$ is the value of the marginal input requirement, or the marginal cost of x_i. Second, if the first two equations of (2.21) are met with equality, the interpretation is that for optimality in the use of b, production of x_i is to be pushed to the margin, x_i^o, where its rising marginal cost just equals its price. On all preceding units, the value to the society of x_i as expressed in p_i is greater than the marginal cost of the unit. On all succeeding units, the marginal cost exceeds the worth of x_i, so it would be irrational for b to be allocated to production of x_i beyond x_i^o.

On the other hand, if the marginal cost of x_i is everywhere greater than its worth, then, rationally, none of the primary resource should be allocated to production of x_i by the agency. This corresponds to the inequality part of the first two expressions in (2.21) and to Plate b of Figure 2.2.

The last expression of (2.21) interprets that if there is some b left over at an optimal X^o, then the shadow price of the resource is zero. The value of another unit of b in creating output when there is already excess b is zero (the expression is met with inequality). But, if there is no b left over (the expression is met with equality), then there is some scarcity value, λ^o, of the resource. That value is the amount by which the value of output would increase if an incremental amount of b were to be found and allocated optimally.

From (2.21) divide the first by the second equation to get

$$\frac{p_1}{p_2} = \frac{\lambda(df^1/dx_1)}{\lambda(df^2/dx_2)}. \tag{2.22}$$

This equation says that in optimality the ratio of the marginal costs of two goods produced in non-zero amounts must be equal to the ratio of their prices. But, the ratio of marginal costs is equal to the marginal rate of transformation of x_2 into x_1. Thus

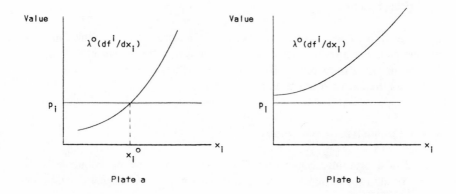

Figure 2-2.

$$\frac{p_1}{p_2} = \text{MRT}_{21}. \qquad (2.23)$$

This is another one of the Pareto conditions that is necessary and together with (2.17) also sufficient, to solve the production allocation problem (2.18).

Now, if we note that p_i is the value of x_i and γ_i is also the value of x_i we can force another equality from (2.14) and (2.23) that is necessary for an optimum allocation,

$$\text{MRS}_{21}^a = \text{MRS}_{21}^b = \text{MRT}_{21}. \qquad (2.24)$$

This equation says that the common rate of substitution of individuals consuming both goods in non-zero amounts must equal the rate of transformation of the two goods that are produced in non-zero amounts.

The planning agency starting from a knowledge of production conditions, f^1, f^2; resource endowment, $\bar{b}$; utility indicators, U^a, U^b; and an ethical judgment about the deservingness of A and B, α^a, α^b, can solve for an allocation of the resource to production, that is a production of x_1^O, x_2^O; a consumption program, x_1^{ao}, x_1^{bo}, x_2^{ao}, x_2^{bo}; prices of the output, γ_1^O, γ_2^O and a price for the input, λ^O.[d] These solution values make welfare in this simple society as great as it can be. The conditions necessary for the solution to obtain are most simply given as the Pareto conditions of Equation (2.24). Together with (2.2) and (2.17), the Pareto conditions are both necessary and sufficient for the welfare optimum. The agency, having chosen solution values for the variables which make (2.24) true, completed its task of allocating resources for a welfare maximum.

Private Maximizing Behavior in a Pricing System

In this section we will examine how successful a price system is in measuring up to the planning agency's performance in allocating resources in an optimal manner in the Pareto sense. The discussion will start with the consumption sector and then move to the production sector.

Private Optimum in Consumption

Instead of a planning agency with ultimate authority over economic decisions trying to allocate resources so as to do best by society, imagine a consuming

---

[d]Note that x_1^O and x_2^O of the production program are the constraints of the consumption program, $\bar{x}_1$ and $\bar{x}_2$. By the same token, the prices γ_1 and γ_2 of output coming out of the consumption program are the prices that must be used in evaluating output in the production program, p_1, p_2. For this reason, the two problems are not solvable recursively for a welfare maximum, but are simultaneously solvable for the welfare optimum.

sector made up of selfish individuals worried only about their own welfare. These consumers spend income on commodities for which they pay prices, the object being to maximize their own well-being. To ease the exposition, let us assume as before that we are dealing with two consumers and two commodities. Each consumer is supposed to have a utility function with which he evaluates various consumption bundles; the function for consumer A is given in (2.1) and the assumptions made about it are given in (2.2). A function with similar properties is attributed to B as well. Assume that y^a and y^b are the income levels distributed to A and B respectively. These levels are clearly arbitrary and reflect an ethical judgment about the deservingness of the individuals. An egalitarian distribution would be $y^a = y^b$. Now, assume that A attempts to

$$\text{maximize: } U^a = U^a(x_1^a, x_2^a) \tag{2.25}$$

$$\text{subject to: } y^a \geqslant p_1 x_1^a + p_2 x_2^a$$

$$x_1^a, x_2^a \geqslant 0.$$

This is a programming problem corresponding to the private optimizing decision. To solve this problem, form the Legrange function[e]

$$L(x_1^a, x_2^a, 1/\alpha) = U(x_1^a, x_2^a) + 1/\alpha[y^a - p_1 x_1^a - p_2 x_2^a] \tag{2.26}$$

The conditions under which this function achieves a saddle point equilibrium are

$$\frac{\partial L}{\partial x_i^a} = \frac{\partial U^a}{\partial x_i^a} - \frac{1}{\alpha} p_i = 0 \qquad x_i^a = x_i^{ao} \tag{2.27}$$

$$< 0 \qquad x_i^a = 0, \quad i = 1,2,$$

$$\frac{\partial L}{\partial (1/\alpha)} = y^a - p_1 x_1^a - p_2 x_2^a = 0 \qquad 1/\alpha = (1/\alpha)^o$$

$$> 0 \qquad 1/\alpha = 0$$

These three equations can, in principle, be solved for x_1^{ao}, x_2^{ao}, $(1/\alpha)^o$, which solves (2.25). These conditions along with (2.2) are necessary and sufficient for these solution values to solve the programming problem. Notice that to get the

[e]The Legrange multiplier in this problem is interpreted as the marginal utility of income. See R. Kuenne, *The Theory of General Economic Equilibrium* (Princeton: Princeton University Press, 1963), p. 70. Because a is the inverse of the marginal utility of income in assigning weights to individual utility in the welfare function, we may write the Legrange multiplier in this problem as $1/a$, rather than introducing new notation.

alpha in $(1/\alpha)^o$ to equal α^a of social optimizing problem, income, y^a, must be chosen properly. Distributing income in this problem is tantamount to assigning ethical weights, α^a, α^b, in the social problem.

Assuming A consumes some of each good, i.e., assuming that the first two equations of (2.27) are met with equality, divide the first by the second equation to get

$$\frac{p_1}{p_2} = \frac{\partial U^a / \partial x_1^a}{\partial U^a / \partial x_2^a} = \text{MRS}_{21}^a . \tag{2.28}$$

That is, subject to the arbitrary, ethical income distribution and prices, individual A maximizes his utility by pushing consumption of each good to the point where its declining marginal utility to him just equals the price he pays. Stated another way, he must equate his marginal rate of substitution to the price ratio.

Of course, if B is a selfish maximizer and has income y^b, then he will attempt to maximize his utility U^b (with property [2.2]) subject to given prices. But, the prices are the same for him as for A. The first order conditions necessary[f] to solve his private consumption programming problem may most easily be stated as

$$\frac{p_1}{p_2} = \text{MRS}_{21}^b \tag{2.29}$$

where the budget constraint is not violated and for goods consumed in non-zero amounts. We can make sure that

$$x_1^{ao} + x_1^{bo} = \bar{x}_1 \tag{2.30}$$

$$x_2^{ao} + x_2^{bo} = \bar{x}_2$$

are the x_i of the social optimizing by choosing the right amount of income to distribute

$$y^a + y^b = y. \tag{2.31}$$

Now, from (2.28) and (2.29) we can get the Pareto conditions for an optimum in consumption,

$$\text{MRS}_{21}^a = \text{MRS}_{21}^b = p_1/p_2 \tag{2.32}$$

---

[f]And with (2.2) also sufficient.

15

Private Optimum in Production

On the production side, replace the benevolent planning agency which allocates a fixed resource to two production activities, with two self-seeking firms. These firms use a primary input to produce x_1 and x_2 respectively with the sole object being to maximize profits. Prices of the input and the outputs are given as λ, p_1, p_2, respectively.

Firm 1 has a profit function

$$\pi^1(x_1) = p_1 x_1 - \lambda f^1(x_1) \tag{2.33}$$

where f^1 is the input requirement function for the firm and λ is the price of input. The assumptions about f are that

$$df^i/dx_i > 0, \tag{2.34}$$

$$d^2 f^i/dx_i^2 > 0.$$

Therefore, λf^1 is the total cost function for firm 1. With prices given, the firm maximizes profits at an output x_1^Q which solves

$$\frac{d\pi^1}{dx_1} = p_1 - \lambda\frac{df^1}{dx_1} = 0 \tag{2.35}$$

or

$$p_1 = \lambda(df^1/dx_1). \tag{2.36}$$

Since the right-hand term of (2.36) is marginal cost, this is a condition requiring that production be pushed to the point where price equals marginal cost for a maximum of profit.

Naturally, if firm 2 buys the primary input at λ and receives p_2 for its output, and if it is a profit maximizer, it will produce at a point x_2^Q where

$$p_2 = \lambda(df^2/dx_2). \tag{2.37}$$

Profits are increasing on all units prior to x_2^Q because marginal cost is less than price, whereas profits would get smaller on all units succeeding x_2^Q because marginal cost is greater than price. The point at which profits are maximized is the point where price just equals marginal cost.

What happens if the amount supplied by firm 1 at p_1 does not equal the amount demanded by A and B at p_1? Then that price will not serve as an equilibrium price in this simple production-consumption economy. Some higher price must be sought with respect to which a maximizing firm will supply more, and with respect to which maximizing consumers will demand less. When a price is found at which

$$x_1^{ao} + x_1^{bo} = x_1^O \qquad (2.38)$$

an equilibrium price has been found. Of course, the same is true of the other market.

But what happens, if at the private maximizing outputs, x_1^O and x_2^O, more of the primary input is demanded than there is to be supplied, $\bar{b}$? Then λ is not an equilibrium price of the input, and a new, higher λ must be sought. This will cause the marginal cost function, $\lambda(df^i/dx_i)$ to shift upward and to the left. That is, the marginal cost of each level of output is now higher because of the higher input price. But, with the greater marginal cost, the profit maximizing output, given output prices, will fall, causing a cut-back in use of b. When the price of the resource has risen to the point where no more is demanded than is available for use, an equilibrium price has been found.

But now, less output is being supplied by the firms, so the output market is out of adjustment. When prices p_1, p_2, λ are found that just equilibrate the input and output markets, these are the ones that should be fed to the private maximizers in order for them to make their optimizing decisions.

From (2.36) and (2.37) notice that we get the condition

$$\frac{p_1}{p_2} = \frac{df^1/dx_1}{df^2/dx_2} = \mathrm{MRT}_{21}. \qquad (2.39)$$

For a private optimum to hold, producers supplying non-zero amounts of output will equate the price ratio to the ratio of marginal costs, or the marginal rate of transformation.

Observing the Pareto condition in consumption, we can see that optimizing firms and consumers will behave so that

$$\mathrm{MRS}_{21}^a = \mathrm{MRS}_{21}^b = \mathrm{MRT}_{21} = p_1/p_2, \qquad (2.40)$$

which are the exact conditions necessary for a social optimum, (2.24). Starting with production conditions, f^1, f^2; utility indicators, U^a, U^b; a fixed amount of primary resource, $\bar{b}$; and an income distribution, y^a, y^b; maximizing producers and consumers will settle on a production program x_1^O, x_2^O, and a consumption program, x_1^{ao}, x_1^{bo}, x_2^{ao}, x_2^{bo}. This economic program will clear all markets if the maximization of all the economic agents is performed with respect to equilibrium prices, p_1, p_2, λ. Also, for appropriate choice of y^a and y^b, marginal utilities of income will result whose reciprocals are equal to α^a and α^b, respectively, the ethical weights given to the consumers by the planning agency.

What we have shown is that the social equilibrium and the private equilibrium are identical for appropriate choices for parameters. The private equilibrium results in a Pareto optimal configuration for the allocation of resources. Moreover, any Pareto optimal allocation of resources, such as that provided by the planning agency, can be achieved by an equilibrium of private maximizers in a competitive setting (i.e., taking prices as given).

3

The Theory of External Effects: Nuisance

Introduction and Definitions

In the last chapter we examined the Neoclassical Welfare Theorem which guarantees Pareto optimality when independent maximizing firms and consumers go about behaving in their own self-interest, subject to (market) price parameters. The question asked in this chapter is, "What happens when firms and consumers are no longer independent, linked only through the market, but when firms and consumers affect one another directly, outside of the market?" That is, what defines an optimum when Arrow's Assumption 2 is violated?

In the literature before 1940, there is a debate about the nature of supply functions.[1] This literature straightens out some confusion on whether rents are or are not social opportunity costs, and whether the supply function truly reflects social marginal costs. This is not the concern here. Rather, this chapter has its logical beginning with the Pigovian discussion of the divergences between the social and private net product. Pigou states that

The source of the general divergences that occur under simple competition is the fact that, in some occupations, a part of the product of a unit of resources consists of something, which, instead of being sold by the investor, is transferred, without gain or loss to him, for the benefit or damage of other people.[2]

Pigou goes on to elaborate on the source of divergences.

Here the essence of the matter is that one person, A, in the course of rendering some service, for which payment is made, to a second person, B, incidentally also renders services or disservices to other persons, C, D, and E, of such a sort that technical considerations prevent payment being extracted from the benefitted parties or compensation being enforced on behalf of the injured parties.[3]

Pigou captures the important points about the nature of externalities in these statements.

Following very clearly in the Pigovian tradition, Viner makes an important distinction between pecuniary and technological externalities.[4] A pecuniary externality is the effect on other firms of an expansion in some firm's output which causes factor prices to increase or decrease for all. In this case, the altered scarcity of factors due to the expansion of some firm is registered through the market in the form of changed factor prices. If firms are maximizers and are free to adjust to the new input prices, then in general a different input ratio, a new set of cost curves, and a different output level (for both firm and industry) results. But if everyone is free to adjust, there is no misallocation. The new equilibrium simply reflects the change in the value of inputs, and the change in the value of inputs is transmitted through the market place in the form of changed prices. For this reason, the external effect is said to be pecuniary.[a] On the other hand, when two decision-making units are linked but "technical considerations prevent payment being extracted from the [affected] parties," there is a technical or, in Viner's words, a technological externality. In this case there is a direct interdependence of utility or cost functions among economic agents which is not registered through the market place.

The definition of a technological externality formulated in mathematical terms by Meade[5] and later by Buchanan and Stubblebine,[6] which lies squarely in the Pigovian tradition, is that a technological externality exists when

$$f^a = f^a(X^a, X^b) \tag{3.1}$$

where X^a is a vector of variables under the control of A and X^b is a vector of variables under the control of other decision-making units. Looking at the consumer case, and letting X^b contain one element, x^b, a marginal externality is said to exist when

$$\text{for } U^a = U^a(X^a, x^b), \text{ it is true that } \partial U^a / \partial x^b \neq 0. \tag{3.2}$$

For $\partial U^a/\partial x^b > 0$ and $\partial U^a/\partial x^b < 0$ there is said to be a marginal external economy and diseconomy respectively.[7] This means that B's variation in x^b affects A. On the other hand, if the externality is pushed so far that $\partial U^a/\partial x^b = 0$ but

$$\int_0^{x^b} \partial U^a / \partial x^b \, dx^b \gtrless 0 \tag{3.3}$$

there is said to be an inframarginal externality. The distinction is this. Equation (3.3) says that adding up all damages (or benefits) caused externally by x^b, there is harm (or benefit) bestowed on A in total by the activity x^b as shown by the fact that the integral is negative (positive). But any additional increment in x^b does not further affect A. This interprets as saying that A is satiated with respect to x^b. If x^b is smoke, this means that there is so much smoke that all crops are

[a]E.J. Mishan, "The Post War Literature on Externalities: An Interpretive Essay," *Journal of Economic Literature* 9 (March 1971): 1-28, feels that to label this an externality is mistaken because it merely describes the working of a market with various decision-making units (firms and consumers) being linked by market interdependence. He does not want the term externality applied to this case. Therefore, applying the label technological externality to the case which follows is a verbal extravagance, according to him, there being no need for the modifier.

dead, all the people are dead, all the property is damaged so that an additional unit of smoke could do no more damage.[8] In Figure 3-1 there is a marginal external economy for $0 \leqslant x^b \leqslant x^b*$, while beyond x^b* there is an inframarginal external economy.

The nature of this nonmarket interdependence is examined by various writers with different emphases emerging. Ellis and Fellner attribute the interdependence to a divorce of scarcity from effective ownership.[9] This divorce means that it is difficult to exclude other firms or individuals from receiving benefits or incurring costs. The full costs or benefits of a given action are not appropriable by the perpetrator of the action through the pricing mechanism. Both Bator[10] and Head[11] recognize that the nonmarket interdependence is in part, at least, due to "ownership externalities." Mishan's contention that externalities arise "...when relevant effects on production or welfare go wholly or partially unpriced,"[12] is the same observation. If the relevant effects were fully priced, then there would be no divorce of scarcity from ownership, and all of the welfare effects would be appropriated by the pricing system. There would be no variable in the cost or utility functions which are under the control of some other decision-making unit because proper exclusion is possible with the effect being priced, or inclusion is possible if the price is paid. But if the price is paid

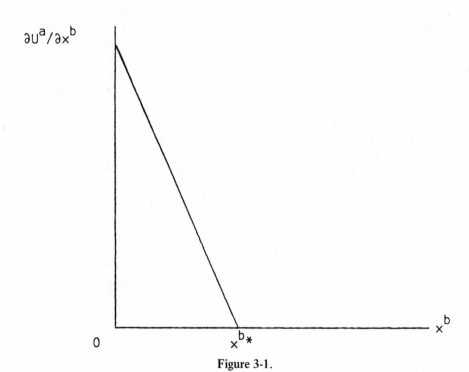

Figure 3-1.

for inclusion, the relevant effect is under the control of the individual or firm doing the purchasing and not some "external" economic unit.

A second theme of emphasis in the discussion of nonmarket interdependence is that it involves joint supply. A related concept is that of public goods. Bator[13] citing Samuelson,[14] Buchanan,[15] and Buchanan and Stubblebine[16] all attribute a joint supply quality to the nature of externalities. If one economic unit takes an action that has welfare effects for other units directly, then there is a jointness in its actions. Relevant welfare effects must be summed over all affected parties to get optimality conditions. By the same token, if there is a quality of publicness to a good, then any action taken with respect to this good affects many other individuals. Again, relevant welfare effects must be summed to get an accurate measure of whether the initial action is economically justified.

But the next question is why, or how, does the jointness and publicness arise? One might be tempted to answer that it is in the nature of certain things to be public goods. Air quality is a public good because my air quality is your air quality and that is just the nature of the problem. On the other hand, a slightly deeper probing might lead us to the idea that air quality is a public good because of the difficulty of appropriability. Exclusion is extremely difficult and therefore relevant effects go wholly or partially unpriced. The two emphases have a relationship. Arrow attempts to unify these approaches by noting that the reason an externality involves joint supply or nonappropriability is that transactions costs which would insure the appropriability of actions are so high that the market is no longer economic.[17] Thus, no market develops and interdependence results.

Mishan makes one important qualification, however. That is, the direct interdependence must be the "... unintentional product of some otherwise legitimate employment,"[18] to qualify as an externality. This qualification is an implication of the Pigovian statement of the externality problem, and it rules out classifying as externalities all kinds of nonmarket interdependence which are intentional. For example, the deliberate sabotage of firms certainly has cost implications and is the result of variables beyond its control. However, it is not the unintentional by-product of some otherwise legitimate economic pursuit, and therefore does not warrant the externality label. Deliberate acts of goodwill or illwill on the part of consumers are not externalities, either, for the same reason.

Moreover, Mishan[19] and Samuelson[20] indicate that there are distinct differences worth maintaining between public goods, joint supply, and externalities. Mishan's classification is the most complete; he derives welfare (necessary) conditions for the cases where there is joint production of private goods, with and without externalities, and joint production of public goods with and without externalities. This classification is complete and more extensive than needed for exposition here.

It seems that there are two main cases which occupy the attention of most writers in the externality area, the nuisance case on the one hand and the pollution and congestion case on the other. They both involve elements of publicness and elements of jointness, so in order to distinguish between these two main cases, let us, following Samuelson, examine the difference between welfare conditions for joint supply and public goods. With individuals displaying a demand (the respective marginal rates of substitution expressed in terms of a numeraire) for two jointly produced goods, the equilibrium is given in Figure 3-2. It depicts the market demand for each good as the horizontal sum of the individual demands. But the worth to the market of the output yielding joint products is the sum of the worth at each argument of the jointly produced goods. That is, the market demands for wool and mutton are added vertically over these jointly produced goods to get the market demand for sheep. Pushing the production of sheep to the margin where this demand equals the marginal cost of sheep production will yield the optimal quantity of sheep. This will, in turn, dictate the supply of wool and mutton which, given the respective demands, will determine a price for wool and mutton. The price will allocate the jointly produced outputs to the respective consumers.

A public good, on the other hand, is a good for which the consumption by one individual does not diminish the supply for others. For example, flood control is a public good because "consumption" of it by one individual in the protected region does not diminish the quantity available to all others in the region. The value, therefore, of flood control is the sum of the values for the respective individuals in the protected region. The individual demand for flood control must be added vertically over individuals to get the market demand. Pushing production of flood control to the margin where this demand equals the marginal cost of flood control yields the optimal quantity which is the same for all individuals.

The first major classification of externalities that is treated in the literature contains elements of jointness and publicness. The central fact about this type of externality is that the legitimate activity and the unintentional harm or benefit occur together directly with no abatement possibilities. For example, Buchanan speaks of a "... 'bad' ... inherent in the output of the industry, an external diseconomy that is directly related to the number of units produced and not to any particulars of the input mix or to the rate of output for any other industry."[21] This is exactly the same conception that Dolbear has in his examination of the externality problem.[22] He assumes that heat is desired for its own sake by one consumer while smoke that is created in the process of obtaining heat is not desired by another consumer. But heat and smoke are created on a one-for-one basis. One unit of heat creates one unit of smoke for appropriately chosen units, and we can in essence talk of only one good, combustion. Another example which serves to illustrate the point is Buchanan and Stubblebine's problem of one property owner building a fence on the border between his property and the adjoining property.[23] Not only does the utility of

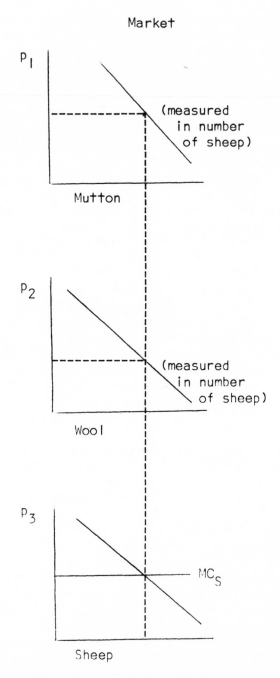

Figure 3-2. Equilibrium for Joint-Supply.

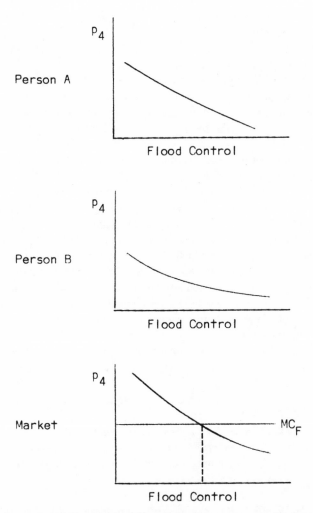

Figure 3-3. Equilibrium for Public Goods. Source: Samuelson, "Diagrammatic Exposition of a Theory of Public Expenditure," *Review of Economics and Statistics* 37 (November 1955): 350-56.

the fence-builder depend on the height of the fence, but also the utility of the resident whose property adjoins the fenced property. The desire for privacy on the part of the builder may unintentionally damage the view of the neighboring resident, with the amount of the damage depending on the height of the fence. It is impossible to have the same height fence with more or less view for the neighbor; that is, it is impossible to abate. All of these cases may be thought of as cases of nuisance. The legitimate activity has associated with it a nuisance to someone else which causes either disutility to consumers or increased costs to firms. Because the activity is inherently a nuisance with no abatement possibilities, the proper economic question is to ask what level of the nuisance-causing activity is optimal. This question always centers on the optimal level of output, if the nuisance is caused by a firm's output, or on the optimal level of consumption, if the nuisance is caused by an individual's consumption. For example, in the case of the fence which poses a nuisance to the neighbor, the question is, "How high *should* the fence be in order to make the sum of the (weighted) utilities of the fence builder and the neighbor as big as possible?"

In terms of Figure 3-4, the nuisance model of externalities can be depicted. The legitimate activity is sheep-raising, and the unintentional, joint, side effect is the waste caused by sheep. To make the exposition simple, assume the amount of waste generated is the same for each sheep and that the waste is somehow inherently damaging like Buchanan's output. Then the sheep cause external marginal damage, the total of which at each level of sheep production is given by the summation of the damage to individuals. The marginal external cost (MEC) is the vertical sum of the individual marginal damage functions because of the public quality of the waste damage. The demand for sheep is the horizontal sum of the individual demands. A private market equilibrium would produce s^1 sheep where all the private market benefits in excess of marginal cost are exhausted. But because of the external cost, sheep production has a social cost of MC_S + MEC, and the optimal output of sheep is s^O where demand equals the marginal social cost.

In view of the above discussion we can formulate the following definitions.

> DEFINITION. *A technological externality exists when the cost function of a firm or the utility function of a consumer is unintentionally dependent on variables under the control of another economic agent (either firm or consumer) as well as variables under its own control.*

> DEFINITION. *A technological externality is defined to be a nuisance when it is not possible to pursue a given level of the legitimate activity with more or less of the unintentional external harm, i.e., when it is impossible to abate.*

The remainder of this chapter explores the allocative implications of the nuisance model.

Optimality in the Nuisance Case

If the technological externality is a nuisance, as is the case with Buchanan and Stubblebine's fence or Dolbear's combustion, then the proper question is, "What

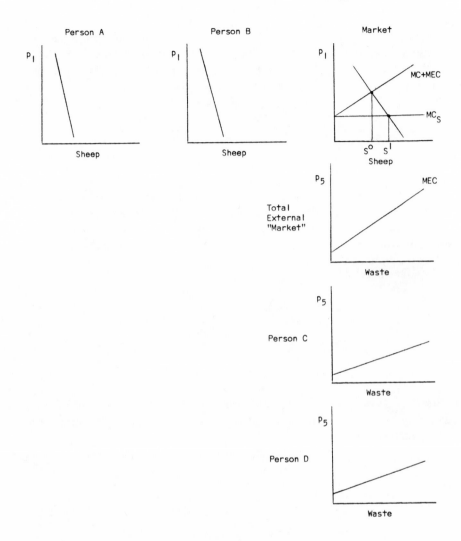

Figure 3-4. Nuisance.

is the optimal level of output which causes the externality?" This section attempts to answer that allocative question.

Model 1. Consumer Affects Consumer
Unilaterally: Nuisance in Consumption

Model and Its Solution. The problem is one of making the society as happy as it can be with the given amount of goods it has. More formally, it is to distribute the goods in such a way that welfare is maximized taking into account that one consumer's actions influence the other consumer's happiness. Specifically, let us suppose there is x_1 and x_2 of goods one and two, respectively, to be distributed throughout a society consisting of two individuals. Their felicity is given by the function

$$U^a = U^a(x_1^a, x_2^a, x_1^b) \tag{3.4}$$

$$U^b = U^b(x_1^b, x_2^b)$$

where U^i is individual i's utility and x_j^i is the ith individual's consumption of good j. We can notice that there is a direct interdependence between the two individuals because of the fact that what person B consumes of good one directly affects person A's utility. The problem stated mathematically is to

$$\text{maximize: } [U^a, U^b] \tag{3.5}$$

$$\text{subject to: } x_1^a + x_1^b \leqslant x_1$$

$$x_2^a + x_2^b \leqslant x_2$$

$$x_1^a, x_1^b, x_2^a, x_2^b \geqslant 0$$

But it is known that a vector maximization problem may be reduced to a problem of maximizing the positively weighted sum of vector elements.[24] The problem then becomes

$$\text{maximize: } W = \alpha^a U^a + \alpha^b U^b \tag{3.6}$$

$$\text{subject to: } x_1^a + x_1^b \leqslant x_1$$

$$x_2^a + x_2^b \leqslant x_2$$

$$x_j^i \geqslant 0, \text{ all } i, j,$$

where $\alpha^a > 0$ and $\alpha^b > 0$. The α^is assign weights to individuals in the welfare function of the society, and Negishi has interpreted them as the reciprocal of the ith consumer's marginal utility of income.[25]

It is necessary for a problem such as (3.6) to have a solution that

$$\frac{\partial^2 W}{(\partial x_j^i)^2} < 0, \qquad i = a, b, \qquad j = 1,2 \tag{3.7}$$

This means that there is universal diminishing marginal utility to consumption of each good. In the case that the externality-creating x_1^b causes consumer A to realize increasing marginal utility, it may not be so severely increasing that it offsets the decreasing marginal utility of x_1^b to consumer A. That is

$$\frac{\partial W}{\partial x_1^b} = \frac{\partial U^a}{\partial x_1^b} + \frac{\partial U^b}{\partial x_1^b} \tag{3.8}$$

must be decreasing in the argument x_1^b for (3.8) to hold.[26]

A problem of the type given in (3.6) is a saddle value problem with a payoff function given by

$$L(x_1^a, x_2^a, x_1^b, x_2^b, \gamma_1, \gamma_2) \tag{3.9}$$

$$= \alpha^a U^a(x_1^a, x_2^a, x_1^b) + \alpha^b U^b(x_1^b, x_2^b)$$

$$+ \gamma_1 [x_1 - x_1^a - x_1^b] + \gamma_2 [x_2 - x_2^a - x_2^b].$$

For an interior optimum to exist, it is necessary that the first partial derivatives of L vanish, or, recognizing the possibility of a corner solution, the complete first order conditions are given as follows:

$$\frac{\partial L}{\partial x_i^a} = \alpha^a \frac{\partial U^a}{\partial x_i^a} - \gamma_i = 0 \qquad \text{for } x_i^a = x_i^{ao} \tag{3.10}$$

$$< 0 \qquad \text{for } x_i^a = 0, \qquad i = 1,2$$

$$\frac{\partial L}{\partial x_1^b} = \alpha^a \frac{\partial U^a}{\partial x_1^b} + \alpha^b \frac{\partial U^b}{\partial x_1^b} - \gamma_1 = 0 \qquad \text{for } x_1^b = x_1^{bo}$$

$$< 0 \qquad \text{for } x_1^b = 0$$

$$\frac{\partial L}{\partial x_2^b} = \alpha^b \frac{\partial U^b}{\partial x_2^b} - \gamma_2 = 0 \qquad \text{for } x_2^b = x_2^{bo}$$

$$< 0 \qquad \text{for } x_2^b = 0$$

$$\frac{\partial L}{\partial \gamma_j} = x_j - x_j^a - x_j^b = 0 \qquad \text{for } \gamma_j = \gamma_j^o$$

$$> 0 \qquad \text{for } \gamma_j = 0,$$

$$j = 1,2.$$

Here is a system of six equations which can be solved for the six variables x_j^i and γ_j. These solution values represent a saddle point of the payoff function and solve the problem of (3.6) for a welfare maximum.

Interpretation. Let X^O and γ^O be the solution vectors that satisfy Equations (3.11). Then we can state the following things about the solution to the optimum problem which occurs at X^O and γ^O. First, either all the goods are consumed (certainly no more can be consumed than exist), or their shadow prices are zero. The reason for this is that if a constrained maximum occurs at X^O for which some amount of a good is left, then adding a tiny bit more of that good to society's initial stock to be distributed would not increase the value of L. The society is then satiated with the good and reducing its scarcity by offering people more of it will not induce them to take more. Its value in making the community better off is zero, and hence its shadow price is zero.

Secondly, consumer A is to press his consumption of each good to the point where his falling marginal evaluation of it equals its price. The same thing holds true of consumer B's consumption of good two. Thirdly, the second equation demonstrates clearly how the technological externality in consumption affects the allocation of x_1. It states that consumer B should pursue consumption of x_1 to the point where its price equals his marginal evaluation of it plus consumer A's marginal evaluation of it. In the absence of this equation being satisfied, too much or too little of x_1 will be consumed by B depending on whether A's marginal evaluation of B's activity is negative or positive, respectively. Either way, if this equation is not satisfied, some other allocation of x_1 would result in a higher level of weighted satisfactions, that is, a higher level of welfare.

Maximizing Behavior and Market Failure. In Chapter 2 we demonstrated that a maximizing individual too small to affect market price will behave so as to equate his own private subjective marginal rate of substitution to the market-determined price ratio,

$$\frac{p_1}{p_2} = \frac{\dfrac{\partial U^a}{\partial x_1}}{\dfrac{\partial U^a}{\partial x_2}} = \frac{MU_1^a}{MU_2^a} = \text{MRS}_{21}^a. \tag{3.11}$$

But individual B will behave exactly the same way if he is a maximizer, thus

$$\frac{p_1}{p_2} = \text{MRS}_{21}^b .$$ (3.12)

Further, we found that equating equals to equals ([3.11] and [3.12]) the rates of substitution of the two consumers are equal if they are parametric maximizers. But this is precisely the institutionally neutral welfare condition that guarantees a Pareto optimal allocation of outputs in consumption when there is no nonmarket interdependence. In the present case where there are externalities in consumption, the individual behavior is not altered. Given the market prices of x_1 and x_2, individuals will still behave by manipulating the variables under their control so that they will achieve a private utility optimum (Equations [3.11] and [3.12]).

However, the social optimum is given by Equations (3.10). Dividing the first by the second equation in (3.10) we get

$$\frac{\gamma_1}{\gamma_2} = \frac{\dfrac{\partial U^a}{\partial x_1^a}}{\dfrac{\partial U^a}{\partial x_2^a}} = \frac{MU_1^a}{MU_2^a} = \text{MRS}_{21}^a .$$ (3.13)

Further, dividing third by the fourth equation in (3.10) we get

$$\frac{\gamma_1}{\gamma_2} = \frac{\dfrac{\partial U^b}{\partial x_1^b}}{\dfrac{\partial U^b}{\partial x_2^b}} + \frac{\alpha^a}{\alpha^b} \frac{\dfrac{\partial U^a}{\partial x_1^b}}{\dfrac{\partial U^b}{\partial x_2^b}} = \text{MRS}_{21}^b + \text{MERS}_{21}$$ (3.14)

Since γ_1/γ_2 are the same (constants) in both (3.13) and (3.14) at the optimum, we can equate equals to equals and get

$$\text{MRS}_{21}^a = \text{MRS}_{21}^b + \text{MERS}_{21} .$$ (3.15)

In this equation, the MRS^i terms represent the ith consumers marginal rate of substituting good two for good one. The MERS_{21} term is the marginal external rate of substituting good two for good one, and it represents A's marginal evaluation of B's consumption of x_1 in terms of B's marginal utility for x_2. It can be interpreted as the amount of x_2 that B would be willing to forego if he had to realize the utility that A *does* realize due to his (B's) own consumptions of x_1.

As stated, private maximizing behavior subject only to market determined price constants (and an arbitrary income distribution) will result in an equilibrium described by the equality of the marginal rates of substitution:

$$MRS_{21}^a = MRS_{21}^b. \qquad (3.16)$$

But from (3.15) we know that a social welfare optimum is described by the equality of A's private marginal rate of substitution (he imposes no externalities on B) with B's social marginal rate of substitution:

$$MRS_{21}^a = MRS_{21}^b + MERS_{21}. \qquad (3.17)$$

B's social rate of substitution is the sum of his private rate and the external rate, and it gives the entire society's (A's and B's) evaluation of B's consumption of the externality causing x_1^b. For example, if $O_A O_B$ is the fixed, total amount of x_1 available to be allocated between A and B (see the constraints of the allocative problem in [3.5]), then we can represent their respective MRS's as the solid lines in Figure 3-5. B's MRS is read from right to left. The lines are solid to represent the fact that they are the functions which A and B "see" and along which they behave privately. The market solution (private maximizing behavior)

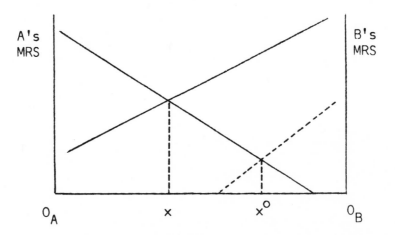

Figure 3-5.

is to allocate $O_A x$ to A and $O_B x$ to B.[b] However, if there is an external diseconomy in B's consumption of x_1, then $\partial U^A / \partial x_1^b < 0$ which implies a negative $MERS_{21}$. Therefore, subtracting $MERS_{21}$ out of B's private MRS_{21}^b we get the social rate of substitution given by the dashed line (because no private individual "sees" it or responds along it). The socially optimum allocation of x_1 is

$$x_1^{ao} = O_A x^o, x_1^{bo} = O_B x^o. \tag{3.18}$$

There is clearly a divergence between the private and social marginal evaluations of x_1^b, which causes a misallocation.

Model 2. Producer Affects Producer
Unilaterally: Nuisance in Production

Model and Its Solution. Suppose there are two production processes which produce output. The output of one of the processes imposes nonabatable costs on the other process, and this nuisance is a function of output only. If we let f^i represent the input requirement function of the ith process, and if there is a fixed amount of a primary input, b, available to both processes, then production conditions can be given as

$$f^1(x_1) + f^2(x_1, x_2) \leqslant b.^c \tag{3.19}$$

The problem now is to allocate b between production of x_1 and x_2 in such a way so as to maximize the value of output. Formally, the problem is to

$$\text{maximize: } V = p_1 x_1 + p_2 x_2 \tag{3.20}$$

$$\text{subject to: } f^1(x_1) + f^2(x_1, x_2) \leqslant b$$

$$x_1, x_2 \geqslant 0,$$

---

[b]Of course, these functions are themselves dependent on the value taken on for x_2 and x_2^{bo}. In actuality, all six variables to the optimum problem are determined simultaneously rather than recursively as the figure would seem to indicate. The figure can only be drawn on the basis that x_2 is allocated "properly."

[c]Very often in the literature, one will see production conditions for such a maximization problem given by an implicit function $F_i(x_1, x_2, b) = 0$, $i = 1,2$, where F_i is the ith production unit's function relating outputs to input. One problem with this formulation is that it obscures the nature of the externality if both firms produce both goods. Moreover, it is felt that using input requirement functions instead of an implicit production function focuses more sharply on the nature of allocating fixed resources to alternative resource-using activities.

where p_i are the market determined prices of output and b is the fixed amount of the primary input.

For purposes of exposition, denote the left-hand side of the constraint in (3.19) as

$$g(x_1, x_2) = f^1(x_1) + f^2(x_1, x_2).$$ (3.21)

We know that for a problem such as (3.20) to have a solution, there must be an increasing marginal input requirement for both production activities, or

$$\frac{\partial^2 g}{(\partial x_1)^2} > 0, \quad \frac{\partial^2 g}{(\partial x_2)^2} > 0.$$ (3.22)

In the case of x_1, this means that

$$\frac{\partial^2 g}{(\partial x_1)^2} = \frac{\partial^2 f^1}{(\partial x_1)^2} + \frac{\partial^2 f^2}{(\partial x_1)^2} > 0.$$ (3.23)

The interpretation is that if activity x_1 causes a declining marginal input requirement in activity two, it cannot be so severely declining that it more than offsets the increasing marginal input requirement in activity one. There must be overall increasing marginal input requirement to activity x_1 for the problem to have a solution.

Assuming that there is enough of the primary input to produce something, we are assured that the problem has a maximum solution. From the Kuhn-Tucker Theorem, we know that the solution must be the optimum value to a saddle value problem with a payoff given by the Legrangian function:

$$L(x_1, x_2, \lambda) = p_1 x_1 + p_2 x_2 + \lambda [b - f^1(x_1) - f^2(x_1, x_2)]$$ (3.24)

The first order conditions necessary for an optimum to exist are

$$\frac{\partial L}{\partial x_1} = p_1 - \lambda \frac{df^1}{dx_1} + \frac{\partial f^2}{\partial x_1} \quad \begin{matrix} = 0 \\ < 0 \end{matrix} \quad \begin{matrix} \text{for } x_1 = x_1^O \\ \text{for } x_1 = 0 \end{matrix}$$ (3.25)

$$\frac{\partial L}{\partial x_2} = p_2 - \lambda \frac{\partial f^2}{\partial x_2} \quad \begin{matrix} = 0 \\ < 0 \end{matrix} \quad \begin{matrix} \text{for } x_2 = x_2^O \\ \text{for } x_2 = 0 \end{matrix}$$

$$\frac{\partial L}{\partial \lambda} = b - f^1 - f^2 \quad \begin{matrix} = 0 \\ > 0 \end{matrix} \quad \begin{matrix} \text{for } \lambda = \lambda^O \\ \text{for } \lambda = 0. \end{matrix}$$

These conditions taken together with the second order conditions (3.22) are both necessary and sufficient for an optimum to exist. If the values which solve (3.25) are denoted as x_1, x_2, λ^O, then it is true that

$$L(x_1, x_2, \lambda^O) \leqslant L(x_1^O, x_2^O, \lambda^O) \leqslant L(x_1^O, x_2^O, \lambda). \tag{3.26}$$

Interpretation. In (3.25) the term of $\partial f^2/\partial x_2$ is the marginal input requirement of activity two. The value λ^O is the optimal scarcity value, or shadow price, of the input b. Therefore, the term $\lambda(\partial f^2/\partial x_2)$ is the value of the marginal input requirement, or the marginal cost. By the same token, the term $\lambda(df^1/dx_1)$ is the marginal private cost of activity x_1 while $\lambda(\partial f^2/\partial x_1)$ is the marginal external cost of x_1. Together, they are the marginal social cost of x_1, or the total resource cost imposed on the economy, both private and external, as a result of pursuing activity x_1. Denoting $\lambda(df^1/dx_1)$ as MPC_1 and $\lambda(\partial f^2/\partial x_1)$ as MEC_1, we can write

$$MSC_1 = MPC_1 + MEC_1. \tag{3.27}$$

In the case of x_2 there are no external costs, so the marginal social cost is just equal to the marginal private cost. Looking at Figure 3-6, we can interpret the first equation in (3.25) as stating that if the marginal social cost of x_1 is not everywhere greater than its price, production should be pushed to the point where its incremental social cost is just covered by price.

Equation two of (3.25) states the same thing for x_2. If x_2 is to be produced at all (i.e., if its marginal social cost is not everywhere greater than its price), production should be pushed to the point where its marginal social cost is just covered by price. But since there are no external costs associated with x_2's production, the marginal social and private costs are equal.[d]

The Legrange function L gives the value of the output plus the value of the unused resources. Naturally, it is desired that the value of the output be as big as possible while the value of the unused resources be as small as possible. Equation (3.26) states that the solution vectors do, in fact, give us this desirable property.

Maximizing Behavior and Market Failure. Dividing equation one of (3.25) by equation two, we get

$$\frac{p_1}{p_2} = \frac{\lambda\left[\dfrac{\partial f^1}{\partial x_1} + \dfrac{\partial f^2}{\partial x_1}\right]}{\lambda\,\dfrac{\partial f^2}{\partial x_2}} = \frac{MPC_1 + MEC_1}{MPC_2} \tag{3.28}$$

---

[d]Since MEC_1 depends on x_2 as well as x_1, Figure 3-6 can only be drawn on the assumption that x_2 is being optimally produced at x_2^O. Actually, x_1, x_2, λ are all determined simultaneously.

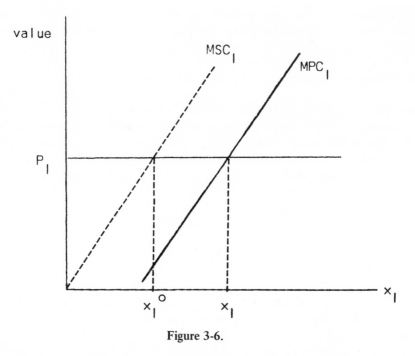

Figure 3-6.

Remembering that the ratio of marginal costs is equal to the marginal rate of transformation, we can write the right hand side of (3.28) as

$$\frac{MPC_1}{MPC_2} + \frac{MEC_1}{MPC_2} = MPRT_{21} + MERT_{21} = MRT_{21} \qquad (3.29)$$

The $MPRT_{21}$, or marginal private rate of transformation, is equal to the rate at which x_1 can be transformed into x_2 counting only the private costs of production. Not producing a small unit of x_1 releases the resources directly needed in production of x_1 to be available to transform into x_2 at a rate depending on its marginal cost. But the $MERT_{21}$, or the marginal external rate of transformation, indicates the rate at which x_1 can be transformed into x_2, counting only the external costs of x_1. Not producing a small unit of x_1 also releases resources which were needed in production of x_2 because of the external effect. These released resources can also be transformed into x_2. The marginal rate of transformation, MRT_{21}, is the sum of these effects. For optimality to occur, it is necessary for the common marginal rates of substitution to be equal to the marginal rate of transformation,

$$MRS_{21}^a = MRS_{21}^b = MRT_{21}. \qquad (3.30)$$

If consumers are maximizers and face market prices, they behave so as to equate their marginal rates of substitution in the absence of nonmarket interdependence. But producers that are linked by technological externalities only "see" and behave along their private marginal cost functions. In Figure 3-6, the producer of x_1 responds only to his own realized marginal costs, MPC_1, which is indicated by the solid line, and ignores the marginal external costs imposed by him on producer two. The decision of how much x_1 to produce should be made taking into account these external costs, or they should be made along the MSC_1 function (dashed line). If the two producers are private maximizers facing market prices and seeking a profit maximum, they produce to the point where price equals marginal private cost, or

$$p_1 = MPC_1, \, p_2 = MPC_2 \tag{3.31}$$

But dividing equals by equals above, we get

$$\frac{p_1}{p_2} = \frac{MPC_1}{MPC_2} = MPRT_{21}. \tag{3.32}$$

Because maximizing consumers equate the output price ratio to their respective MRS's, we know that

$$MRS_{21}^a = MRS_{21}^b = MPRT_{21}. \tag{3.33}$$

But this is not the equality that must hold for Pareto optimality. The external rate of transformation is ignored by market maximizers and their behavior violates one of the Pareto conditions. There is a divergence between the private cost and the social cost which results in the failure of private maximizing behavior to solve the welfare optimum problem of (3.20). Referring again to Figure 3-6, we can see that private maximizing behavior in general leads to more x_1 relative to x_2 being produced than is optimal.

Model 3. Producer Affects Consumer Unilaterally

Model and Its Solution. The last of the nuisance models we will explore in which there are unilateral nonmarket effects is the model in which the producer affects the consumer. The simplest model that captures the allocative significance of this problem is one with two productive processes and one consumer. Let x_1 and x_2 be produced according to the input requirement functions

$$f^1(x_1) = b^1, \, f^2(x_2) = b^2 \tag{3.34}$$

where there are increasing marginal input requirements to both processes,

$$d^2 f^1 / dx_1^2, > 0, \ d^2 f^2 / dx_2^2 > 0. \tag{3.35}$$

The consumer evaluates consumption bundles according to

$$U = U(x_1, x_2, x_1^p) \tag{3.36}$$

where the x_1^p represents the fact that a nuisance is associated with the production of x_1 (hence the mnemonic p), and the nuisance enters the utility function of the consumer directly, without the intervention of a market. The nuisance may be in the form of a by-product associated with the production of x_1, but if it is, there exists no possibility of abatement of the by-product. In Buchanan's words, the "... 'bad' ... [is] inherent in the output of the industry, an external economy that is directly related to the number of units produced and not to any particulars of the input mix or to the rate of output for any other industry."[27] The utility function has the following properties.

$$\partial U / \partial x_2 > 0, \ \partial^2 U / \partial x_2^2 < 0 \tag{3.37}$$

$$\frac{\partial U}{\partial x_1} + \frac{\partial U}{\partial x_1^p} > 0, \ \frac{\partial^2 U}{\partial x_1^2} + \frac{\partial^2 U}{(\partial x_1^p)^2} < 0$$

$$\frac{\partial U}{\partial x_1^p} < 0.$$

That is, there is overall diminishing marginal utility to both goods. The externality may not be of such a nature that second order assumptions are violated.

If there is a fixed amount of the primary input, $\bar{b}$, then the problem is one of allocating $\bar{b}$ between the alternative production processes in order to make the consumer as happy as possible. That is

$$\text{maximize: } U = U(x_1, x_2, x_1^p) \tag{3.38}$$

$$\text{subject to: } f^1(x_1) + f_2(x_2) \leqslant \bar{b}$$

$$x_1, x_2 \geqslant 0.$$

Again, we know a problem like this has a solution if there is enough of the primary input to produce something. The solution is the optimum value to the saddle value problem with the payoff function given by the Legrangian:

$$L(x_1, x_2, \lambda) = U(x_1, x_2, x_1^p) + \lambda [\bar{b} - f^1(x_1) - f^2(x_2)]. \quad (3.39)$$

The full first order conditions necessary of an optimum to exist taking into account the possibility of a corner solution are

$$\frac{\partial L}{\partial x_1} = \frac{\partial U}{\partial x_1} + \frac{\partial U}{\partial x_1^p} - \lambda \frac{df^1}{dx_1} = 0 \qquad x_1 = x_1^o \qquad\qquad (3.40)$$
$$< 0 \qquad x_1 = 0$$

$$\frac{\partial L}{\partial x_2} = \frac{\partial U}{\partial x_2} - \lambda \frac{df^2}{dx_2} = 0 \qquad x_2 = x_2^o$$
$$< 0 \qquad x_2 = 0$$

$$\frac{\partial L}{\partial \lambda} = b - f^1 - f^2 = 0 \qquad \lambda = \lambda^o$$
$$> 0 \qquad \lambda = 0$$

These conditions when taken together with the second order information of (3.35) and (3.37) are both necessary and sufficient for an optimum to exist. The values which solve (3.40), $X^o = [x_1^o, x_2^o]$ and λ^o, are such that, according to the Kuhn-Tucker Theorem,

$$L(X, \lambda^o) \leqslant L(X^o, \lambda^o) \leqslant L(X^o, \lambda). \qquad (3.41)$$

Interpretation. The final equation of (3.40) guarantees either that the fixed input is exhausted or that its price is zero. The second equation states that the production of x_2 be pushed to the point where its rising marginal cost equals the falling marginal evaluation of it by the consumer. If the marginal cost is

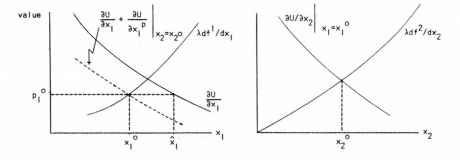

Figure 3-7.

everywhere greater than the consumer's marginal evaluation of it, then x_2 is not produced. The first equation gives the same kind of marginal-evaluation-equals-marginal-cost condition for the optimal use of b in production of x_2 that equation two gives for production of x_1. The difference is that the marginal disutility of the nuisance caused by production of x_1 must be included in the marginal evaluation.

It is interesting to note that in the case of one consumer and one firm, the external disutility caused by the nuisance can be completely internalized by the market. If the firm producing x_1 were instructed to sell all the output it could, provided only that it charged marginal cost,[e] then the market would clear at x_1^O and the firm would charge p_1^O. But this is the optimal quantity of x_1 that a benevolent planning agency would wish to have produced. The reason that a market solution yields a social optimum with only one consumer is that he can accurately translate all of marginal disutility due to the nuisance into market demand information. At p^O he demands only x_1^O even though he would demand $\hat{x}_1$ in the absence of x_1 being a nuisance to him. There is no free-rider problem to contend with when there is only one consumer.

Extending Model 3. If, however, the problem is extended to include two or more consumers the solution is not so neat. For simplicity, let there be two productive processes, one of which causes a nuisance to each of two consumers. The problem then is to allocate the input, b, to the two productive processes and then to allocate the output to the two consumers so as to maximize welfare. Symbolically, we wish to

optimize:

$$L(x_1^a, x_2^a, x_1^b, x_2^b, \lambda) = \alpha^a U^a(x_1^a, x_2^a, x_1^p) \qquad (3.42)$$
$$+ \alpha^b U^b(x_1^b, x_2^b, x_1^p) + \lambda[\bar{b} - f^1(x_1) - f^2(x_2)].$$

The conditions necessary for an optimum are

$$\frac{\partial L}{\partial x_1^a} = \alpha^a \frac{\partial U^a}{\partial x_1^a} + \sum_{j=a,b} \alpha^j \frac{\partial U^j}{\partial x_1^p} - \lambda \frac{df^1}{dx_1} = 0 \qquad x_1^a = x_1^{ao} \qquad (3.43)$$
$$< 0 \qquad x_1^a = 0$$

$$\frac{\partial L}{\partial x_1^b} = \alpha^b \frac{\partial U^b}{\partial x_1^b} + \sum_{j=a,b} \alpha^j \frac{\partial U^j}{\partial x_1^p} - \lambda \frac{df^1}{dx_1} = 0 \qquad x_1^b = x_1^{bo}$$
$$< 0 \qquad x_1^b = 0$$

---

[e]This is necessary so the single firm will not exploit its monopoly position. We could think of the marginal cost function as a competitive industry marginal cost, or supply, function and thus eliminate the problem of monopoly power. The problem here is to explore the optimum solution, not the allocative effects of market structure.

$$\frac{\partial L}{\partial x_2^j} = \alpha^j \frac{\partial U^j}{\partial x_2^j} - \lambda \frac{df^2}{dx_2} = 0 \qquad\qquad x_2^j = x_2^{jo}$$
$$< 0 \qquad\qquad x_2^j = 0, \quad j = a,b.$$

$$\frac{\partial L}{\partial \lambda} = \bar{b} - f^1 - f^2 = 0 \qquad\qquad \lambda = \lambda^o$$
$$> 0 \qquad\qquad \lambda = 0$$

The optimality conditions for consumption and production of good x_2, which is not a nuisance, are the same as before. Each consumer pushes his consumption to the point where his individual marginal evaluation of x_2 equals the common marginal cost of production.

The nuisance good is another matter. Not only is the x_1 demanded by A a nuisance to him, but it is also a nuisance to B as well. The social evaluation of A's consumption must include the disutility caused to all members of society who are affected by the nuisance. There is some publicness to the good x_1, though it is not purely a public good. In consumption, the good is strictly private, but yet production causes consumers disutility because of some annoying but nonabatable aspect of production which is solely a function of the quantity produced. In this respect, the good has a public quality to it.

Comparing this result to the model with only one consumer, we can see that the social evaluation of A's consumption cannot totally be translated into market demand information solely by A's own behavior. If we let B represent "everyone else," the reason for this can be seen. In a market situation where selling of x_1 must take place at marginal cost, A has no incentive to restrict his demand for x_1 to reflect his own marginal disutility for the nuisance aspect of the good. This is because of the fact that he still has to bear the disutility from "everyone else's" demand. The free-rider syndrome exerts its powerful influence on behavior and prevents a market situation from arriving at a socially optimal allocation.

Separability and Nonseparability in Nuisance

The issue of separability is most often treated in the context of two firms, one causing a nuisance to the other.[28] But the issue extends also to the pollution and the congestion cases[29] and will be treated in that context in Chapter 4. By separable cost functions, we mean a function with the property that

$$C(x_1, x_2) = c_1(x_1) + c_2(x_2). \qquad (3.44)$$

In the case of our Model 2, the input requirement function for firm 2, the object of the nuisance, is

$$f^2(x_1, x_2) = b^2 \qquad (3.45)$$

But λ, the shadow price of the input, times the required input will yield the value, or cost, of the required input. Therefore, the cost function for firm 2 in this model is

$$C^2(x_1, x_2) = \lambda f^2(x_1, x_2) \tag{3.46}$$

and if the technological externality is separable, then

$$C^2(x_1, x_2) = \lambda [f_1^2(x_1) + f_2^2(x_2)]. \tag{3.47}$$

From this we can formulate the definition of separability.

> *DEFINITION. A nuisance is said to be separable when variations in the nuisance do not affect the marginal cost (utility) of the offended party. That is, a separable nuisance exists when*
>
> $$\partial^2 C^2(x_1, x_2)/\partial x_1 \partial x_2 = 0$$
>
> *where C^2 is the cost (utility) function of the offended party.*

In practical terms this means that x_1, the nuisance output, causes firm 2 to incur costs which increases both its total and average costs. But the incremental costs of production remain unchanged. That is, the new total cost function at a higher level of x_1 is merely a vertical displacement of the old cost function at a lower level of x_1. It does *not* mean, however, that no additional costs are incurred as x_1 increases; additional costs may well be incurred which have the effect that the total and average costs in the production of x_2 are higher. But the marginal cost of production of x_2 remains unaltered.

On the other hand, a nonseparable nuisance is one where variations in the

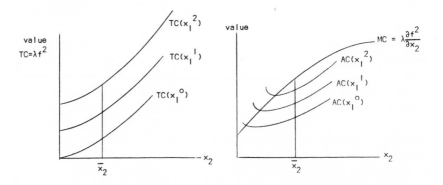

Figure 3-8. The Effect of Separability on the Offended Party.

nuisance output cause variations in the marginal cost of offended firm's output. Thus $\partial^2 C^2(x_1,x_2)/\partial x_1 \partial x_2 \neq 0$, which means that it is not possible to "separate" the cost function as it is in (3.47),

$$C^2(x_1, x_2) \neq \lambda [f_1^2(x_1) + f_2^2(x_2)]. \tag{3.48}$$

This leads to the definition of a nonseparable nuisance.

> *DEFINITION. A nuisance is said to be nonseparable when variations in the nuisance affect the marginal cost (utility) of the offended party. That is, a nonseparable nuisance exists when*
>
> $$\partial^2 C^2(x_1, x_2)/\partial x_1 \partial x_2 \neq 0$$
>
> *where C^2 is the cost (utility) function of the offended party.*

Not only is there a whole family of total and average cost functions, one for each level of x_1, but there is a corresponding family of marginal cost curves as well.

Allocative Implications of Separability

The optimality conditions of Equations (3.25) are mathematically valid conditions for the two producer nuisance model without regard to separability. But, these conditions have allocative implications that are dependent on the separability issue. Let us graphically examine an interior equilibrium given by (3.25) with all equations met by equality. If firm 1 is forced by a planning agency mandate to produce at the optimal output, x_1^O, and if firm 2 is allowed to operate within a perfect market, then x_2 is produced at x_2^O and unit costs are just covered, $AC(x_1^O)$. Now, suppose that the planning agency adopts a "hands-off" policy toward firm 1. It now ignores social costs and is concerned only with private costs. The cost structure to which it is responsive is represented by the solid curves in Figure 3-9, plate a. The firm now adjusts output to the point where the marginal private cost equals output price at $\overline{x}_1$.

What happens to firm 2 if the externality is separable? The marginal cost function remains unchanged as does the output price, p_2. However, an expansion of x_1 to $\overline{x}_1$ causes firm 2 to realize higher unit costs because of the externality. Its unit cost function shifts to $AC(\overline{x}_1)$ causing it to realize a deteriorating profit position. In fact, because at the old (x_1^O, x_2^O) equilibrium firm 2 was just realizing normal profits, its position has changed to one of loss. The firm will make no marginal adjustments because its output price and marginal cost are the same regardless of how much x_1 is produced. Only its

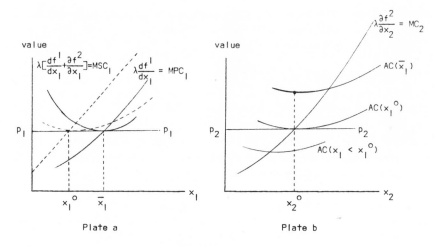

Figure 3-9.

average cost and therefore its profit picture is affected by the increase in x_1. But, if the firm expects its average cost increase to be a permanent, long-run phenomenon, i.e., if it does not expect x_1 to cut back production to x_1^O or below, it will close down. The increase in production of the nuisance-creating x_1 does not affect the marginal decision of the offended firm if costs are separable, but it can affect the total decision, i.e., the decision whether or not to produce at all. If the offended firm remains in production after the offending firm expands output past the optimal level, that implies that excess profits are being earned by firm 2.

PROPOSITION. A unilateral, separable nuisance in production affects the total, but not the marginal decision of the offended firm.

The Severity of the Nuisance. We know the way in which firm 2 requires additional input as a function of the nuisance-creating output is given by

$$f^2(x_1, x_2) = b^2 \tag{3.49}$$

from (3.20).

But, because the nuisance is separable, we can write (3.39) as

$$b^2 = f_1^2(x_1) + f_2^2(x_2) \tag{3.50}$$

The marginal cost of producing x_2 is

$$MC_2 = \lambda df_2^2(x_2)/dx_2 \tag{3.51}$$

and can be seen to be clearly independent of x_1. However, the social cost of producing x_1 is given by

$$C_1 = \lambda [f^1(x_1) + f_1^2(x_1)] \tag{3.52}$$

with a marginal social cost of production equalling

$$\text{MSC}_1 = \lambda [df^1/dx_1 + df_1^2/dx_1]. \tag{3.53}$$

The nuisance may require firm 2 to have an incremental need for input which is constant with respect to x_1 (the dotted line), decreasing with respect to x_1 (dashed line), or increasing with respect to x_1 (dotted and dashed line). The marginal social cost, accordingly, differs from the marginal private cost by a constant, decreasing, or increasing amount depending on the severity of the nuisance.

Allocative Implications of Nonseparability

The general optimum conditions (3.25) for the two firm model are valid for the case of nonseparable cost functions as well as for the separable case. There are unique allocative implications, however, that are dependent on this issue. As before, the solid lines represent functions which are "seen" by the respective private decisionmakers and along which they respond. The output of firm 1 causes a nuisance to firm 2 and therefore carries with it a cost external to the

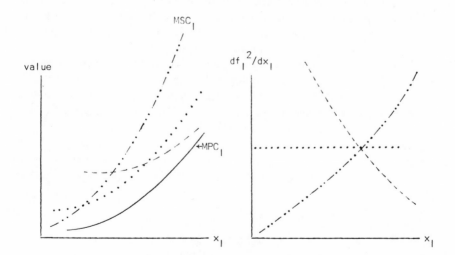

Figure 3-10.

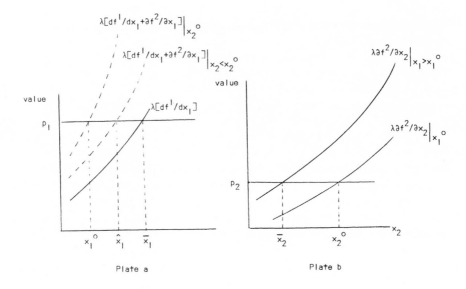

Figure 3-11.

firm. Assume that there is no market interference, and that the two firms reach a private equilibrium. Firm 1 does not realize the externality as an opportunity cost, so it equates the marginal private cost (solid curve, Plate a, Figure 3-11) to the market determined price at $\overline{x}_1$. This level of the nuisance-causing x_1 will determine a relevant marginal private cost for firm 2. Since the nuisance is nonseparable, the level of x_1 determines the marginal cost of x_2 with increases in x_1 increasing the marginal cost. Let the uppermost curve (Plate b, Figure 3-11) represent firm 2's marginal cost at $x_1 = \overline{x}_1$. Then firm 2 pushes output to the point where its marginal cost, which includes private costs and externally imposed costs, equals the market price of x_2. This occurs at $\overline{x}_2$.

The private equilibrium $(\overline{x}_1, \overline{x}_2)$ is the solution which maximizes individual profits, but it is not the social optimum which affords a joint maximum of profits. Suppose that this externality problem comes under the jurisdiction of a planning agency whose sole function is to economize. Realizing that the $(\overline{x}_1, \overline{x}_2)$ solution does not afford the largest social product, it will attempt to modify the behavior of the two firms in a desirable direction.[f] Assume by some appropriate means the agency could get firm 1 to realize the external marginal cost it imposes on firm 2 as an opportunity cost. At the level of production $(\overline{x}_1, \overline{x}_2)$ this is

$$\text{MEC} = \lambda \partial f^2 (\overline{x}_1, \overline{x}_2) / \partial x_1. \tag{3.54}$$

---

[f]The following discussion is not for the purpose of proposing policy, but merely for the purpose of bringing to light the nature of nonseparable cost functions.

When this external cost is imposed, firm 1's marginal cost is no longer equal to the solid line in Plate a, Figure 3-11, but rather it is equal to a new marginal cost function represented by the middle line (which is dotted). Accordingly it adjusts output from $\overline{x}_1$ to $\hat{x}_1$. But when it does this, it changes the marginal cost of firm 2 which drops at each argument,

$$\lambda \partial f^2 (\hat{x}_1, x_2)/\partial x_2 < \lambda \partial f^2 (\overline{x}_1, x_2)/\partial x_2. \qquad (3.55)$$

This occurs because less output from firm 1 means less nuisance to firm 2 at each argument. This drop in firm 2's marginal cost means it will expand output. The larger output, however, from firm 2 alters the marginal external cost of firm 1's output. If the agency is successful in devising a method of getting firm 1 to realize this new, higher marginal external cost, the output of firm 1 falls even further. This, in turn, alters the marginal cost of x_2 and it continues to expand output. This process continues until a point is reached where the resources saved by a further reduction in x_1 plus those saved by reduction of the external cost would be more than taken up by the expansion of x_2. At this point (x_1^0, x_2^0) a joint maximum, social optimum is achieved. There still exists an external marginal cost measured by $\partial f^2(x_1^0, x_2^0)/\partial x_1$, but there is no longer any Pareto-relevant external cost.

In the nonseparable case, the marginal decision of the offended firm depends on the level of output of the offending firm. It is also true that the marginal external cost depends upon the level of output of both firms.

PROPOSITION. A unilateral, nonseparable nuisance in production affects the marginal decision of the offended firm.

The Severity of the Nuisance. The external marginal cost of production of x_1 is

$$\text{MEC} = \lambda \partial f^2 (x_1, x_2)/\partial x_1. \qquad (3.56)$$

In the discussion of nuisance it is clear that $\partial f^2/\partial x_1 > 0$, that is, production of x_1 causes production of x_2 to require more resources. But given x_2, increases in x_1 can make this external marginal cost either increase, remain constant, or decrease, depending on the severity and nature of the nuisance. For example, if the marginal external cost is positive and increasing, i.e.,

$$\partial^2 f^2/\partial x_1^2 > 0 \qquad (3.57)$$

then the difference between the marginal social and marginal private costs of x_1 widens with respect to x_1, and increases in x_1 cause the marginal cost curve of x_2 to shift upward and to the left. However if

$$\partial^2 f^2 / \partial x_1^2 < 0, \tag{3.58}$$

then increases in x_1 cause the gap between marginal social and marginal private costs of x_1 to decrease.

Bilateral Effects

It is true that any one of the models explored above may be modified to include reciprocal, or bilateral, external effects. It is also true that the bilateral effects in any one case, say the producer case (Model 2 above), may be separable, nonseparable, or a combination of the two. For example, if

$$C^1 = C^1(x_1, x_2), \ C^2 = C^2(x_1, x_2) \tag{3.59}$$

represent the cost functions of the two producers in Model 2, it may be that both C^1 and C^2 are nonseparable, or separable, or it may be that one of the functions is separable and the other is not. The same observation holds for the other models as well. It is interesting to note that very often nonseparability in the nuisance case is discussed in the context of bilateral effects.[30] In fact, it is not hard to conclude reading the literature that bilateral effects and nonseparability are related.[31] That, in fact, is not the case. The issue of bilateral, or multilateral, effects is logically distinct from the issue of separability.

Direction of the Misallocation

As is noted earlier in the chapter, if an output creates a nuisance in production to another producer the social cost at the private market equilibrium is greater than the private cost. It seems natural to suppose that the quantity produced at a private market equilibrium would be too great relative to what it should be at a Pareto optimal configuration. There are, however, income effects which must be accounted for. As the economy moves from a suboptimal equilibrium to an optimal one, real income and output is increased, which causes an income effect that may work in the same or opposite direction from the effect of substituting all other goods for the nuisance good in production. The relative strengths of these effects if they are in opposite directions will determine whether more or less of the nuisance good is produced in an optimal configuration.

But Buchanan and Kafoglis show a counter-example in which there are no income effects and yet the aggregate quantity of a good generating an external economy (i.e., displaying a "positive" nuisance) is smaller at the optimum than at a suboptimal competitive equilibrium.[32] This unorthodox result is the same as finding an example of good which affords a negative nuisance to someone else of which it is true that a larger amount is required in an optimal allocation than in a suboptimally competitive one. This unorthodox result is disturbing and is the subject of investigation by several writers. The question that presents itself is whether there are so many exceptions to the orthodox rule that the rule is in danger of becoming the exception. In all cases the writers examining this problem assume reciprocal (bilateral, in the case of two decision units) externalities, although Buchanan and Kafoglis demonstrate the unorthodox result for unilateral externalities as well.

Vincent finds that a necessary condition that the unorthodox result hold is that the second order conditions for competitive solution (where decision-units behave as individual maximizers) be violated in the case of one party to the externality.[33] He also finds that it is necessary for the unorthodox case that "... one party's [consumption] be a more-than-perfect substitute for the other's [consumption]."[34] Baumol finds a counter-example in which the unorthodox result holds even if second order conditions are not violated.[35] This example, however, involves the circumstance of the marginal and total externalities going in the opposite direction.[g] Schall presents a set of necessary conditions for the orthodox case to hold.[36] They are that the primary (priced) factor of production be identically distributed to all (two) firms which are a party to the externality and that all (both) firms that are party to the externality exert equal marginal externalities on the others.[37]

In an excellent paper by Diamond and Mirrlees the question is further addressed as to whether the orthodox result, which seems intuitively correct, is violated by so many exceptions that it should be discarded, or whether the unorthodox result is an anomalous exception with such a rare chance of occurrence that it can be ignored.[38] To broach this question, Diamond and Mirrlees look for a set of sufficient conditions that will rule out the unorthodox result, but, since all results fall in either the orthodox or unorthodox category, the sufficient conditions they seek will guarantee that the orthodox result holds. If such conditions are found, they are much more useful than necessary conditions, because knowledge that sufficient conditions obtain insures that the unorthodox result does not hold. The same cannot be said, of course, for knowledge of necessary conditions.

For the two-person, constant supply-price case involving nuisance in consumption, Diamond and Mirrlees find that certain partial derivate information and a "normality" assumption meet their needs. In particular, for two consumers, A and B, with utility functions, $U^a = U^a(x^a, x^b)$ and $U^b = U^b(x^a, x^b)$, it is sufficient to rule out the unorthodox case that (1) each consumer prefer his

---

[g]An externality influences only the total and average costs (utility) if the function is separable while the marginal cost (utility) is changed as well if the function is nonseparable. The Baumol example involves an externality which exerts external *marginal economies*, but external *total diseconomies*. See W.J. Baumol, "External Economies and Second Order Optimality Conditions," *American Economic Review*, 54 (June 1964): 368.

own consumption to that of the other person on the margin, (2) that utility functions be quasi-concave, and (3) the externality be "normal." In the case of external diseconomies, the normality assumption means that A is willing to abide increases in the nuisance only if compensated with more income (numeraire); that is, A's marginal rate of substitution of the nuisance good for everything else is positive.[39]

In fact, as can be seen from the following two theorems by Diamond and Mirrlees, a subset of three conditions presented above is enough to rule out the unorthodox case.

THEOREM. If

$$(1) \quad \frac{\partial^2 U^a}{(\partial x^a)^2} \leqslant \frac{\partial^2 U^a}{\partial x^b \partial x^a}$$

when $\partial U^a / \partial x^a = p_x$ *(constant supply price), and*

$$(2) \quad \frac{\partial^2 U^b}{(\partial x^b)^2} \leqslant \frac{\partial^2 U^b}{\partial x^a \partial x^b}$$

when $\partial U^b / \partial x^b = p_{x'}$

then the anomalous case cannot occur for either economies or diseconomies.[40]

This says that condition (2) above, or quasi-concavity of the utility functions, is alone sufficient to rule out the unorthodox case.

THEOREM. If

$$(1) \quad \partial U^a / \partial x^a > \partial U^a / \partial x^b, \textit{ and}$$

$$(2) \quad \partial U^b / \partial x^b > \partial U^b / \partial x^a$$

and if the externality is normal as defined above, then the anomalous case cannot occur.[41]

This theorem says that condition (1) and (3) above, or normality and the marginal preference for own-consumption, are alone sufficient to rule out the unorthodox case.[h]

---

[h]Buchanan and Kafoglis' as well as Vincent's "more-than-perfect substitute" condition is a denial of the preference of own-consumption.

An interesting observation, and one which Diamond and Mirrlees do not explicitly make but which flows from their first theorem, is the following:

COROLLARY. If U^a *and* U^b *are separable, the unorthodox result is ruled out.*

> *Proof: The definition of separability is that*
>
> $$\frac{\partial^2 U^a}{\partial x^b \partial x^a} = \frac{\partial^2 U^b}{\partial x^a \partial x^b} = 0.$$
>
> *If own-consumption displays the standard property that*
>
> $$\frac{\partial^2 U^j}{(\partial x^j)^2} < 0, \; j = a,b$$
>
> *then, the conditional clause of the first theorem must hold.*

Diamond and Mirrlees extend their results to the case of many consumers. In this case as well, restriction on derivates and deductions therefrom furnish the basis of theorems which rule out the unorthodox case. For the case of a changing supply price, consider these three conditions: (1) the competitive demand slopes downward, (2) the demand in an optimal configuration slopes downward, and (3) the unorthodox case is impossible not just at the competitive price (p_x of the first theorem), but at any fixed price. Then the following theorem is true for the variable supply price case:

THEOREM. If the anomalous case does not occur at a constant competitive supply price, then it is sufficient for the anomalous case not to occur with a nondecreasing supply curve that any two of the three above conditions hold.[42]

With this set of theorems, we have the stronger sufficient conditions for the orthodox case to hold.

While it may not be possible to completely vindicate the economist's intuition about the direction of a move from a suboptimal competitive equilibrium to a socially optimal Pareto equilibrium when either a positive or negative nuisance exists, it is the case that assumptions commonly made by economists concerning the nature of utility and demand are sufficient to insure that less, not more, of a nuisance is required at optimum than at a competitive equilibrium. This is not to rule out as a very real possibility the case of Kafoglis and Buchanan where less "health care" is required optimally (due to collectivization) than is provided in a competitive, private setting. But, if this case holds, it

52

means that at least one of the three Diamond-Mirrlees conditions are violated or that demand curves slope upward.

Market Structure and Nuisance

The standard critique of noncompetitive forms of market organization is that private maximizing behavior under those market structures will yield a solution in which price is greater than marginal cost. That means that resources are used elsewhere in the economy to purposes which are less valuable in the eyes of consumers than they would be in production of the output of the noncompetitive industry. There is therefore no Pareto optimum in an economy characterized by noncompetitive market forms. With reference to Figure 3-12, a competitive industry structure with marginal cost equal to p_c produces at x_c. On the other hand a monopoly produces at x_m and charges p_m.

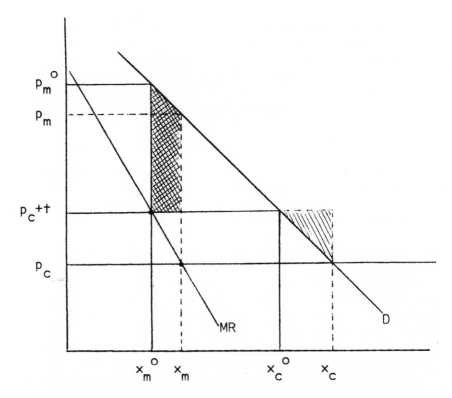

Figure 3-12. Source: J.M. Buchanan, "External Diseconomies, Corrective Taxes, and Market Structure," *American Economic Review* 59 (March 1969); 175. Copyrighted 1969 by American Economic Association.

Suppose the output x imposes a nuisance somewhere in the economy of t per unit of output. Then imposing a tax of t per unit of output would cause the competitor to restrict output to x_c^O with a welfare gain equal to the striped region in Figure 3-12. That is, the welfare gain is the reduction in damages, $t(x_c - x_c^O)$, minus the loss of consumer's surplus, which is the area under the demand curve net of marginal cost between x_c^O and x_c. On the other hand, if the industry is monopolistic, the imposition of a tax in the amount of t per unit would result in output being restricted from x_m to x_m^O. The resulting change in welfare is actually a welfare loss equal to the cross-hatched region in Figure 3-12. This represents the loss in consumer's surplus net of the damages saved.[43] The point is clear. Imperfect markets cause the wrong quantity to be produced, while a nuisance also causes the wrong quantity to be produced. But, in the nuisance case, correcting the quantity produced cannot be properly undertaken without reference to both the market structure and the externality, since misallocations due to both sources hinge on the level of output.

The policy implications must be carefully drawn. For example, it is tempting to relegate every smoke-producing monopoly to this category. The example of steel mills in Gary, Indiana, comes to mind. An argument might go as follows: A suboptimally small amount of steel is produced, *ceteris paribus,* because of its oligopolistic market structure. But because smoke from the steel plants imposes an external cost, a suboptimally large amount of steel is produced, *ceteris paribus*. Perhaps the two effects are mutually offsetting so that nothing should be done. But this line of reasoning is faulty. The Buchanan model deals with the nuisance case while the Gary steel case is clearly a case of pollution with abatement possibilities. The important thing to note may be summarized in the following proposition.

> *PROPOSITION. If any producers in the economy are linked by technological externalities of the nuisance type in an imperfectly competitive setting, then there are two sources of misallocation. The first is misallocation due to market structure, and the second is misallocation due to the nuisance. Policy made with regard to only one or the other of these problems will in general not afford Pareto improvements in welfare.*

On the other hand, the models explored in this chapter treat price parametrically. The results of that exploration might be summarized in the following proposition.

> *PROPOSITION. If any decision-making units are linked by a nuisance, and if all units are small relative to the market so that the prices are parameters, then profit- and utility-maximizing behavior implies that the necessary conditions for a welfare maximum fail to obtain.*

Several points need to be made. First Arrow shows that for an appropriate redefinition of the commodity space to provide a distinct dimension for the effect of a good on distinct individuals, externalities are defined away.[44] Goods that appear in any one individual's utility function do not appear in anyone else's. Attached to this set of goods is a price vector which supports an optimal allocation. As Arrow points out, the above proposition (for commodities defined in the standard way) is true not so much because markets fail as because markets fail to exist. There is not a close enough alignment of the scarcity involved with transfer rights that can be priced. This is primarily because the transactions costs necessary to cause a closer alignment are very high. Markets therefore do not appear.

The empirical question posed by Demsetz[4,5] and implicit in Arrow is whether the transactions costs incurred in securing allocative gains left unexploited by the market are greater or less than those gains. The theoretical question examined in the policy chapters is what effects alternative policies may be expected to have on allocation.

4

The Theory of External Effects: Pollution and Congestion

Introduction and Definition

The second major classification of externalities which involves elements of publicness and jointness is the pollution and congestion type. While pollution and congestion may appear to be different phenomena, Rothenberg provides the basis for treating them together. Pure congestion, according to Rothenberg, occurs when users generate equal quality deterioration per unit of activity and share equally in resulting lower quality, while pure pollution occurs when some users generate high levels of quality impairment and others generate none, whereas the latter group suffers the entire quality reduction. The general case is where all users both generate and share in the damage, but at various rates.[1] The central fact about this type of externality is that the legitimate activity can be pursued at a given level with either more or less of the unintended harm resulting. This is because of the possibility of abatement which is presented if the externality is of the pollution or congestion type. This leads us to the following definition.

> *DEFINITION. A technological externality is defined to be of the pollution or congestion type when it is possible to pursue a given level of the legitimate activity with more or less of the unintentional external harm, i.e., when it is possible to abate.*

In this model of externalities, the theoretical focus shifts away from the question of the proper level of the legitimate activity toward the question of the proper level of external activity (for example, smoke). If car air pollution can be abated, the question is not how many cars should we have on the road, or even how many miles should we permit the cars to drive, but rather how much smoke should we permit to enter the air.

Basic Model

The emphasis on jointness can be misleading in the case of pollution.[a] While we might be tempted to think of smoke as being jointly produced with steel, a more

[a]Because of Rothenberg's synthesis, the use of the word "pollution" wherever it appears may be substituted with the word "congestion" without loss of economic meaning. While there are unique considerations in congestion problems as distinct from pollution problems, these considerations are no different than the special considerations warranted in air versus water pollution problems.

correct way to view pollution in terms of air use is that it is an input into the production process. By analogy, we may imagine labor disutility as being jointly produced with output, but it seems to be more nearly correct in terms of labor use to treat labor as an input, with production taking place with either more or less labor. If we think of production taking place with the two traditional inputs, capital and labor, and one environmental input, the absorption of air measured in units of smoke discharged, we get a production function

$$x = f(k, l, s). \tag{4.1}$$

Provided that $\partial f/\partial s \neq 0$ we can write an inverse of (4.1) as

$$s = s(k, l, x) \tag{4.2}$$

If f is representative of an economy-wide, aggregate production function, then it is clear that there are other ways of changing the quantity of pollution besides simply changing the level of output. On the consumption side, analogous observations hold. If we view consumption not as consumption of things but as consumption of services,[2] then it is clear that there are substitution possibilities in providing those services.

If we treat all uses of k and l which substitute for s as "abatement" and label these activities A, then a basic feature of the model is the function (4.3)

$$s = s(x, A) \tag{4.3}$$

where

$$\partial s/\partial x > 0, \ \partial s/\partial A < 0. \tag{4.4}$$

The other primary components of the model are a function evaluating the damage caused by s and a function detailing the cost of reducing s. Let the damages function be

$$D = D(s) \tag{4.5}$$

where

$$dD/ds > 0 \text{ and } d^2D/ds^2 > 0, \ 0 < s < \overline{s}. \tag{4.6}$$

This means that more pollution causes more damage, but at an ever increasing rate, at least up to some point. One might suppose that at least after some point dD/ds slopes downward, meaning that more pollution causes more damage at a decreasing rate.[3] These causes are depicted in Figure 4-1.

Let the function

$$B = B(s) \tag{4.7}$$

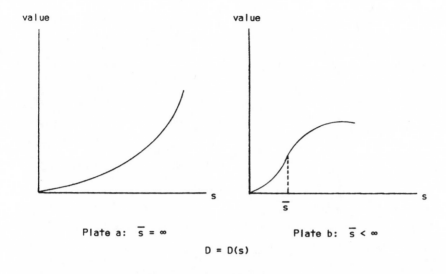

Plate a: $\bar{s} = \infty$　　　　Plate b: $\bar{s} < \infty$

$D = D(s)$

Figure 4-1.

be the benefits from avoided abatement costs which is dependent on the level of pollution. Because the costs of abatement rise more slowly with the first units of pollution abated than with the last, the benefits function will have a slope such that

$$dB/ds > 0 \text{ and } d^2B/ds^2 < 0. \tag{4.8}$$

This reflects the increasing marginal cost of abatement. To demonstrate more clearly the relationship of D and B, refer to Figure 4-2. With reference to the s axis, D is the total damage due to the pollution level s. The function B gives the costs avoided in permitting pollution level s to exist. If the figure is turned upside-down so the functions are read with respect to the s-abated axis, we see that B is the total cost of abating. Costs increase at an increasing rate and the marginal cost of abating the last unit of s is very high indeed. On the other hand, D is the total damage avoided due to abating at a given level.

In deciding what is the optimal level of pollution (or pollution abatement) that should take place, one must trade these two costs against one another; that is, the cost of pollution damage is traded against the cost of abatement. The object is to minimize the true cost of pollution which is the sum of these costs, or to maximize the difference between abatement costs avoided, or benefits, and the pollution costs. The problem stated mathematically is

$$\text{maximize: } \pi(s) = B(s) - D(s). \tag{4.9}$$

The social profit is maximized (or the social costs are minimized) when

58

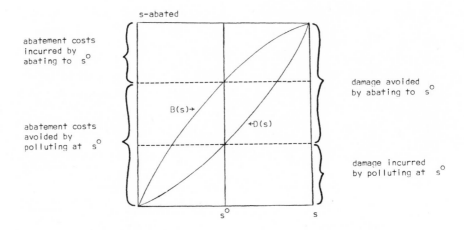

Figure 4-2.

$$d\pi/ds = dB/ds - dD/ds = 0 \qquad\qquad (4.10)$$

or

$$dB/ds = dD/ds \qquad\qquad (4.11)$$

which may be solved for s^O, the optimal level of pollution.

Figure 4-3 shows this equilibrium graphically and the interpretation is as follows. If one unit less pollution than s^O were required, the marginal cost of abating that unit (dB/ds) would be larger than the marginal reduction in damages (dD/ds). Therefore, that level of pollution would be suboptimally small, and the sum of abatement costs and pollution damages would be larger than necessary. The converse is true if one unit greater than s^O were allowed. In this case, the marginal cost of abating another unit of pollution is less than the damage caused by that unit of pollution. Clearly, abating another unit of pollution and avoiding the damage would reduce the overall costs of pollution. At $s^O + 1$ units of pollution, there exists a suboptimally large amount of pollution or a suboptimally small amount of abatement.

What happens, however, if the damages function is like that depicted in Plate b of Figure 4-1? That is, at some high level of pollution, damage increases at a decreasing rate as almost everything that can be damaged is in fact damaged. There are two cases: The case depicted in Plate a and Plate b of Figure 4-4, respectively. In either case, the slope functions would appear as depicted in Plate c. Examining Plate c we find that there are two points at which condition (4.11) holds. Which is the optimal point?

Taking first the situation in Plate a, we find that at s^1 the total damages

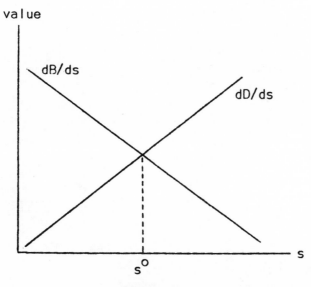

Figure 4-3.

exceed the total abatement costs avoided. At s^1 a point of maximum social loss due to pollution is reached, the loss being $L = D(s^1) - B(s^1)$. In the second case presented in Plate b, the first order condition (4.11) is also met at two points. At point s^2 there is still some net social profit, $\pi = B - D$, which is as small as it can get at s^2. At s^O this difference is clearly as large as possible; s^O is a maximum and s^2 is a minimum of social profit. The condition (4.11) is therefore only necessary for the optimal level of pollution to be achieved. A condition when taken together with (4.11) that is both necessary and sufficient for an optimal level of pollution is

$$d^2\pi/ds^2 < 0 \text{ which implies } d^2B/ds^2 < d^2D/ds^2. \qquad (4.12)$$

This means that once pollution has proceeded to the point where damages are increasing only at a slower and slower rate, we must be careful to require that the incremental costs avoided (benefits) are increasing at an even slower rate. Otherwise, second order conditions are violated, and theorems regarding the optimality of tax and bargaining solutions are invalid.[4]

Interpretation of Optimum as a
Quasi-Market Equilibrium

Referring to Figure 4-5, the function dB/ds is the marginal expense not incurred by permitting that last unit of pollution, reading with respect to the s axis (left

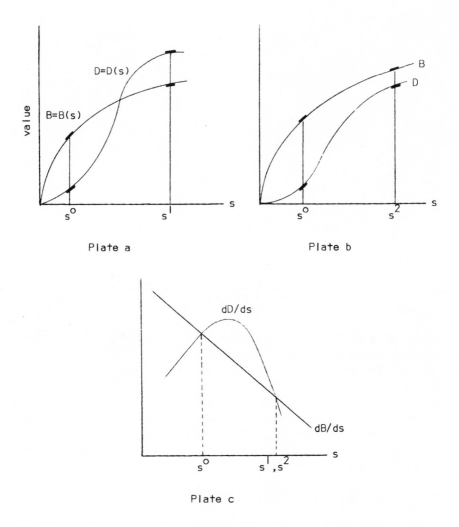

Plate a

Plate b

Plate c

Figure 4-4.

to right). However, with respect to the *s*-abated axis (reading from right to left) *dB/ds* is the marginal cost of abatement. At each argument, *dB/ds* gives the maximum amount the polluter would offer to be permitted to expel that unit of pollution rather than abate it. As such, the marginal benefit function *dB/ds*, may be interpreted as a demand function. Although there is no market for pollution, the demand-like characteristics of the function are present reflecting increasing marginal cost of abatement, and for these reasons the function may be termed a quasi-demand function for pollution. If the ordinate value that the function takes on is measured not in dollar terms, but rather in terms of some numeraire

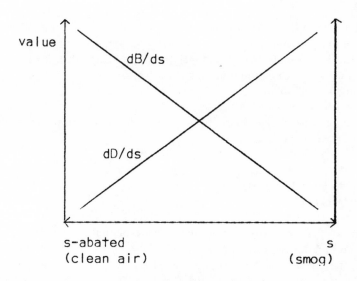

Figure 4-5.

good, then at each argument, s, the function dB/ds represents the marginal rate of substitution of the numeraire good for pollution. It is what the polluter would be willing to give up in terms of the numeraire to be permitted to issue that unit of pollution rather than abate it.

On the other hand, dD/ds gives the marginal damage imposed by the last unit of pollution. It gives the minimum amount that those realizing the damage would be willing to accept in order to tolerate the marginal unit of pollution, and therefore has the same interpretation as a market supply function. But there being no market for pollution, dD/ds cannot be said to be a supply function. If the value axis is measured not in terms of dollars, but rather in terms of a numeraire good, the quasi-supply function gives the rate at which the numeraire good can be "transformed" into pollution. That is, at each argument dD/ds gives the marginal rate of transformation of the numeraire into pollution.

We are now in a position to interpret Figure 4-3 as if it were a market equilibrium; that is, as a quasi-market equilibrium. If, for simplicity, we assume one decisionmaker is doing the polluting and one decisionmaker is being harmed, the necessary conditions (4.11) for a socially optimal level of pollution which minimizes its total costs are

$$\text{MRS}_{xs} = \text{MRT}_{xs}. \tag{4.13}$$

But from (2.24), this condition is just the Pareto condition that must hold for a welfare maximum.

We can go one step further in the market analogy. Just as supply and demand forces equilibrate on a market to determine an equilibrium price and quantity, so the quasi-market demand and supply for pollution determine a value for pollution at the cost-minimizing, optimal level of s. This value given by the ordinate value of the equilibrium in Figure 4-3 measures precisely the opportunity cost of a unit of pollution. If one unit of pollution were foregone, this value is a measure of the resources that would be released. Its interpretation is exactly that of a competitive market price, but because no market exists, we must call it a quasi-market price.

Pollution with Many Pollutees:
The Public Quality of Pollution

Discussion of the case where more than one economic unit suffers from pollution necessitates the further interpretation of Figure 4-3. In the same way that the total functions, D and B, can be interpreted with respect to s or s-abated, so the marginal functions, dD/ds and dB/ds, can be interpreted with respect to s or s-abated. To put a little more content into the discussion, we may think of s as smog (air pollution of a stylized, homogeneous type), and s-abated as clean air. Naturally, the more smog that is abated, the cleaner the air.

With this in mind, let us take the interpretation of these functions with respect to clean air and ask what the quasi-market demand for clean air is if there is more than one individual sharing the air quality. If there are two people sharing air quality, the quasi-market demand for air quality (clean air) is the vertical sum of their individual demands as noted in Figure 3-3. The individual demands are dictated by what damage is avoided at each given quality level, and the sum of these demands is their willingness to pay for a given quality level. As noted before, in real terms, this demand is their marginal rate of substitution. So quality level s^O is valued in the amount v^{ao} and v^{bo} by A and B respectively, and by the community, then, in the amount of the sum of the individual valuations.

The optimal air quality is now not met when the conditions (2.24) for an optimum of a private good are met. Rather, the optimum for a public good occurs where the sum of the marginal rates of substitution equals the marginal rate of transformation, as noted in Figure 3-3.[5] The discussion of the interpretation of dD/ds and dB/ds with respect to either the s axis or the s-abated axis brings to light another observation. The demand for pollution is identical to the supply of clean air (abatement); they are both dB/ds. Moreover, the supply of pollution is identical to the demand for clean air (abatement); they are both dD/ds. The interpretation of dB/ds and dD/ds depends on whether they are being read with respect to the s axis or the s-abated axis.

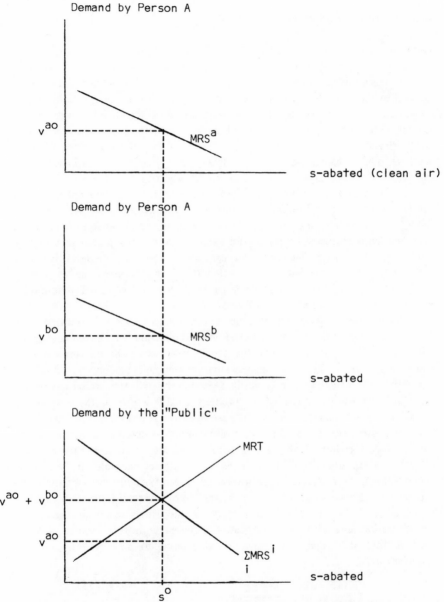

Figure 4-6. Source: Samuelson, "The Pure Theory of Public Expenditure," *The Review of Economics and Statistics* 36 (November 1954): 387-89; "Diagrammatic Exposition of a Theory of Public Expenditure," *The Review of Economics and Statistics* 37 (November 1955): 350-56.

Market Area and Optimality in Pollution

Definition

As with market phenomena, the quasi-market supply and demand schedules may be used to elucidate the optimal level of pollution under changing conditions. To make the analysis more realistic, we must say what the "market area" is. In dealing with a private good with no externalities, the supply and demand refer to some relevant market area. For example, we refer to regional markets, national markets, and even world markets. Where the factors that affect market conditions (demand and supply) vary from one region to the next, but not within the region, a market is said to be regional. As an example, housing appears to be a regional market because the factors that affect supply, primarily labor, raw materials, and land costs, vary from region to region. The factors affecting demand, primarily population density and per capita income, also vary from region to region. Therefore, we can talk of the San Francisco Bay Area housing market and the San Joaquin Valley (California) housing market because of the interregional variations in the price (and quantity) of identical homes on identical-sized lots with identical services.

When we depict quasi-market demand and supply curves for pollution as in Figure 4-5, what kind of a quasi-market are we referring to? The demand varies with the polluting activity while the supply varies with the damage resulting from the pollution. Therefore, the quasi-market area must be constructed so that it contains all of the supply effects (damage) due the pollution sources within it. Such an area has been termed a "problem shed," which in the case of air pollution is an "airshed," and in the case of water pollution is a "watershed." Following our line of reasoning, it is appropriate to extend the notion to the case of a "congestion shed," as well. While there are almost certainly spillover effects between adjacent problem sheds (especially for example, in the case of air pollution), these effects are of the second order of magnitude compared with the primary effects occurring within the shed. For this reason the problem shed is treated as if it is "airtight" with respect to other sheds and it emerges as the proper market area with which to deal. The fundamental fact about a problem shed is that it is highly unlikely to correspond to any existing political boundaries.

The Nature of the Problem Shed

In the case of pollution the problem shed is not a well-defined deterministic shed. Rather, it is stochastic in nature. For example, in the case of air pollution, meteorological events shape the airshed in differing ways. The well-known inversion (temperature) layer that often blankets the Los Angeles basin prevents a vertical mixing of air and effectively puts a low elevation lid on the air space that is available to the region. But the inversion layer is not a deterministic meterological event that occurs with certainty every day of the year, or a given number of days a year. Rather it is a stochastic event that occurs according to a probability distribution. Moreover, inversion conditions vary in intensity; some are almost impenetrable, while others are weak. This means that the same volume of airborne effluents creates varying degrees of damage based on the airshed which is stochastically determined by meteorological events. The same observation may be made with regard to stream flow in the case of water pollution.

There are differences in air and water pollution based primarily on differences in the nature of the problem shed. First, it is easier to control hydrological events than meteorological events.[6] This provides a range of alternatives on water pollution abatement that is not available in air pollution abatement. Next, because of the receiving medium, water treatment can take place before consumption of the water in general. This option is not a major alternative in air pollution, abatement of which is primarily centered on treatment processes which prevent airborne effluents from ever getting into the air.[7]

Comparative Statics and Optimality in the Pollution Level

If there is a shed[b] with a given demand D_1 and supply S of pollution (Figure 4-7), then the socially optimum, cost minimizing level of pollution in the basin occurs at s_1. The opportunity cost of a unit of pollution at that level is v_1 reflecting the incremental damage and the incremental cost of abating.[c] Assume that for whatever reason new pollution generating activity is drawn to this basin with no change in the population or its damage potential. This means that the demand conditions have changed, but the supply conditions have not. Specifically, demand for pollution has shifted to D_2 reflecting that at each level of s the polluters would be willing to pay more to pollute rather than to abate that marginal unit. The fact that the population and its damage potential remain unchanged means that the supply schedule for pollution is constant. The equilibration of these two forces means that the opportunity value of a unit of pollution has risen to v_2 from v_1 because the cost of abating ($s_2 - s_1$) exceeds the damage inflicted. It is rational, then, from society's point of view, to permit s_2 level of pollution rather than maintain the old s_1 level.

Plate c depicts the situation where demand conditions in a basin are constant and where supply conditions have changed, for example, as when new residents move into an air basin. In this case, the demand function is constant but the

b"Basin" and "region" are used interchangeably with "shed."
cAt the optimum, these two are equal.

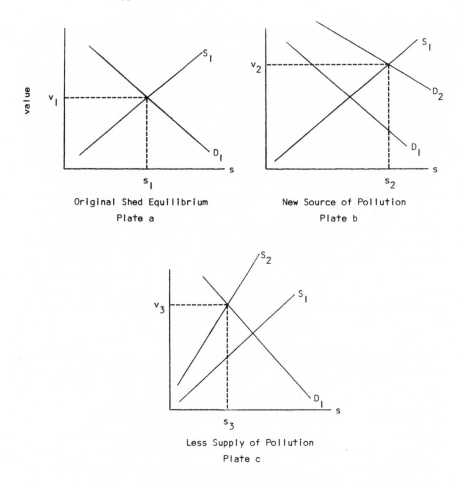

Original Shed Equilibrium
Plate a

New Source of Pollution
Plate b

Less Supply of Pollution
Plate c

Figure 4-7.

supply function shifts from S_1 to S_2 reflecting the fact that an incremental puff of smoke causes more damage. In this case the equilibration of quasi-supply and demand forces results in a higher opportunity cost of pollution, v_3, and a lower optimal level of pollution, s_3, being established. On all units of pollution between s_1 and s_3 it is cheaper to abate than to incur the damage, so the cost minimizing level of pollution decreases typically with new residents moving into an air basin with a constant demand for pollution.

Because the supply curve reflects incremental damage due to pollution, there is a component of the supply curve which reflects things that have value but cannot be translated precisely into dollars and cents worth of damage. For example, in the case of air pollution, the monotonous greyness and loss of visibility is damage. By the same token, the foul smell of severely polluted water is damage apart from the tangible loss of fish and recreational opportunities. But what is the loss of these amenities worth? Damage to aesthetic values, while very real physically, is nevertheless intangible economically. In a purely theoretical sense, it should be possible to develop some willingness-to-pay concept (not to have the smell, for example) by asking everyone who is affected by the aesthetic loss what it would be worth to them not to have the aesthetic loss.[d] The main point is that both tangible and intangible damage must be included in the supply function for the quasi-market equilibrium to represent a social optimum.

A final observation centers on the fact that demand for pollution depends on abatement technology. If and when new and more inexpensive waste treatment processes are found, the demand for pollution at each price is reduced. Or, stated another way, the maximum amount a demander of pollution is willing to pay rather than abate a given marginal unit is less after the new technology is available. One might also suspect that the supply of pollution, especially the intangible part, varies with income. The expectation is that the greyness of the polluted atmosphere weights more heavily in the evaluation of damage of the high, rather than the low, income population. One implication is that we might expect the supply of pollution in a given shed to experience a leftward secular "drift" in response to growth, other things equal.

Least Cost Reduction in Pollution

One Pollution Source

For simplicity, assume that there is only one pollution source in the basin under consideration. Then the function (4.3) gives the way in which pollution is "produced" as a function of both the level of production and the level of abatement. After establishing that the optimal level of pollution in the basin is s^O (see Figure 4-3), how can the polluter reduce the flow of pollution to s^O at a minimum of cost?

We can plot that information contained in Equation (4.3) by letting X be the number of units of x withheld from production. The curves in Figure 4-8 represent equal levels of pollution reduction, with curves further to the right denoting lower levels of pollution. The fact that they intersect the axes indicates that with no abatement, reductions in output will reduce s by given amounts, while with no reductions in output, increments in abatement will reduce s by given amounts. The curvature of the isopollution-reduction curves represents the fact that there are diminishing (marginal) returns to at least the abatement activity. That is,

---

[d]While literal implementation of this questionnaire method would be prohibitively expensive in most cases, an estimate could be provided by sampling techniques. The free-rider problem emerges again, however. See Chapter 5.

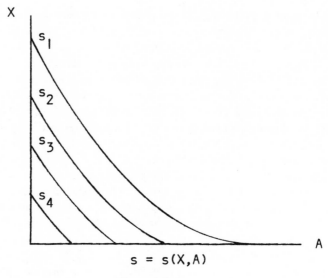

$$s = s(X,A)$$

Figure 4-8.

$$\partial^2 s/\partial A^2 < 0. \tag{4.14}$$

There might be either diminishing or constant marginal returns in pollution reduction to cutting back output, but as long as there are diminishing returns to abatement, the curves will have the shape depicted.

The cost of reducing x by an amount X and of using an amount A of abatement activity is

$$C = p_x X + p_a A \tag{4.15}$$

where p_x is the given output price and p_a is the given unit price of abatement activity. Equation (4.15) is a line in (X,A) space which interprets as all the pairs of (X,A) that can be purchased for the same cost, C, at constant prices.

Formally, the problem of minimizing the costs of pollution reduction is to

$$\text{minimize: } C = p_x X + p_a A \tag{4.16}$$

$$\text{subject to: } s^0 - s(X, A) = 0$$

$$X, A \geqslant 0.$$

Mathematically, we may do this by forming the Legrange function and setting all first partial derivatives equal to zero,

$$L(X, A, \mu) = p_x X + p_a A + \mu [s^0 - s(X, A)] \tag{4.17}$$

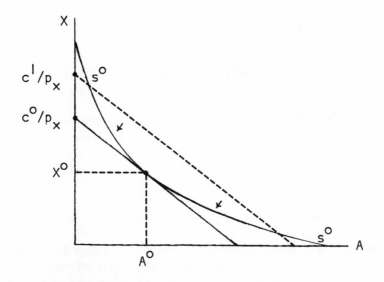

Figure 4-9.

$$\partial L/\partial X \;=\; p_x \;-\; \mu \partial s/\partial X \;\;= 0 \qquad\qquad\qquad (4.18)$$

$$\partial L/\partial A \;=\; p_a \;-\; \mu \, \partial s/\partial A \;\;= 0$$

$$\partial L/\partial \mu \;=\; s^O \;-\; s(X, A) \;\;= 0.$$

Dividing the second equation of (4.18) by the first, transposing and cross-multiplying, we get that

$$\frac{\partial s/\partial X}{p_x} = \frac{\partial s/\partial A}{p_a} \qquad\qquad\qquad (4.19)$$

when

$$s^O \;-\; s(X, A) \;=\; 0. \qquad\qquad\qquad (4.20)$$

That is, the minimum cost pollution-reduction policy (X^O, A^O) must be one where the per dollar marginal productivity in reducing s must be the same for both reducing alternatives. Were it not so, a change in policy could reduce costs even further by allocating away from the alternative with the lower toward the one with the higher per dollar marginal reduction capability. At (X^O, A^O) it is true that the cost of the abatement policy given by

$$C^O = p_x X^O + p_a A^O \qquad (4.21)$$

is as low as it can be and still bring the pollution level to s^O.

There might be cases in which the marginal rate of technical substitution is either everywhere greater or everywhere less than the ratio of the prices of the policy alternatives. In these cases, the optimal policy is a corner solution, relying on only one of the policy alternatives. In the case depicted in Plate a of Figure 4-10, the marginal rate of technical substitution is everywhere smaller in absolute value than the slope of the isocost line, implying a cost minimizing strategy of $(X^O, 0)$.

Many Pollution Sources,

Suppose there are for simplicity two sources of pollution with respective demand for pollution curves, dB^1/ds_1 and dB^2/ds_2. These functions reflect the marginal cost of abatement of the two producers when they are read from right to left. In Figure 4-11, firm 1's demand for pollution is relatively inelastic as compared to firm 2, which means its marginal cost of abatement is more rapidly rising. These marginal abatement cost functions represent the minimum cost for which each marginal unit can be abated.

If unbridled pollution is permitted, these two firms would pollute at the level $\bar{s}$. But the cost minimizing level of pollution is given by s^O, the equilibrium of a quasi-market supply of and demand for pollution. The supply function is the vertical sum of the marginal damage functions of all individuals in the basin,

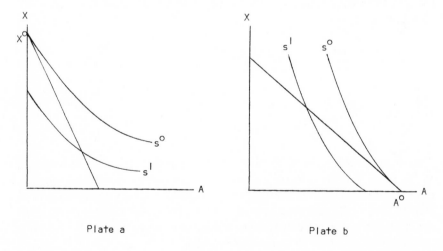

Plate a   Plate b

Figure 4-10.

71

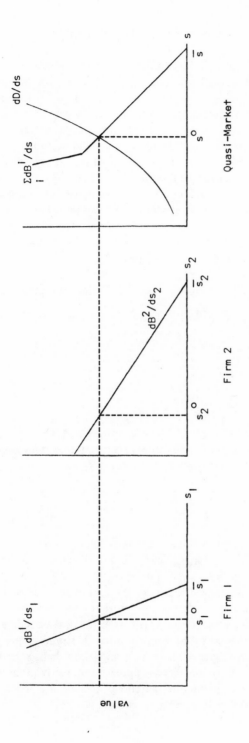

Figure 4-11. Interfirm Efficiency in Pollution Reduction.

while the demand is the horizontal sum of the individual polluter's demand functions.

In order for the level s^O of pollution to be achieved at a minimum of cost, it must certainly be true that whatever level of pollution reduction each firm pursues, it does so by employing a minimum cost abatement policy (see previous section). That is, there must certainly be intrafirm efficiency. But with more than one source of pollution, it must also be true that whatever aggregate amount of reduction is optimal, the distribution of the pollution reduction among firms is such as to take best advantage of the marginal cost differentials across firms. That is, there must also be interfirm efficiency.

Analytically, we can derive interfirm efficiency conditions by requiring that the abatement costs avoided minus the pollution damage be as large as possible. That is, we wish to maximize

$$\pi(s_1, s_2) = B^1(s_1) + B^2(s_2) - D(s)^e \tag{4.22}$$

with respect to s_1 and s_2. The conditions necessary for this maximization are

$$dB^1/ds_1 - dD/ds = 0 \tag{4.23}$$

$$dB^2/ds_2 - dD/ds = 0$$

or

$$dB^1/ds_1 = dB^2/ds_2 = dD/ds. \tag{4.24}$$

This equilibrium is depicted in Figure 4-11 as occurring at (s_1^O, s_2^O). The firm for which abatement is easier (costs rise more slowly) is required to abate more than the firm for which abatement is harder ($\bar{s}_2 - s_2^O > \bar{s}_1 - s_1^O$).

Other Dimensions

As Turvey[8] and Crocker[9] point out, there are other relevant dimensions in the discharge of effluents besides the quantity per period of time (how much). There are also dimensions of where and when. For example, smoke discharged may cause less damage if it is discharged higher in the atmosphere, or waterborne waste may be less damaging if it is discharged further down the river. By the same token, whether the discharge is made evenly over the time period or occurs all at once makes a difference as to the damage incurred. Moreover the precise time of the day (or month or year) also has an impact on damage costs. The point made by Crocker and Turvey is that to approach optimality in these other dimensions, a change should be made if the cost of making it is less than the incremental damage cost avoided in making the change.

$^e s = s_1 + s_2$

Separability in the Damage Functions

The concept of separability extends to the damage functions due to pollution as well. If the damage from two independent sources of pollution cause damage which is just equal to the sum of the damages from each source taken alone, then the damage function is said to be separable. If, however, the various pollution sources interact, then the total damage is not merely the sum of the damages due to each source taken alone. This leads to the following definition.

DEFINITION. For s_1 and s_2 two independent sources of pollution, if the total damage function can be written

$$D(s_1, s_2) = D(s_1) + D(s_2),$$

then the damage function is said to be separable. That is the damage function is separable if $\partial^2 D(s_1, s_2)/\partial s_1 \partial s_2 = 0$. On the other hand, if $\partial^2 D(s_1, s_2)/\partial s_1 \partial s_2 \neq 0$, then the damage function is nonseparable and cannot be written as a sum.

For example, two independent sources of pollution, "generalized" air pollution and smoking, seem to be nonseparably related to mortality rates (see Table 4-1). If we take a rural dwelling (no air pollution), nonsmoking population and wish to note the effect that moving to the city and smoking has on its mortality rates, we could proceed by assuming the effects were separable. Then by adding 127, which is the independent effect of smoking on death rates in a rural population, and 33, which is the independent effect of moving to the city on death rates in a nonsmoker, to 11, the rate of a nonsmoking rural population, we get a mortality rate of 171. But the actual mortality rate is 226. There are 55 deaths per 100,000 population unaccounted for if the damages are separable. The conclusion must be[f] that the extra 55 deaths per 100,000 population occur as the result of the interaction of smoking and air pollution. The damage

Table 4-1
Mortality Rates per 100,000 Man Years

	Rural	Urban	Difference
Smoker	138	226	88
Nonsmoker	11	44	33
Difference	127		55

Source: L.B. Lave, "Air Pollution Damage: Some Difficulties in Estimating the Value of Abatement," in A.V. Kneese and B.T. Bower (eds.) *Environmental Quality Analysis* (Baltimore: Johns Hopkins Press for Resources for the Future, 1972), p. 217-18.

[f]Must be, that is, if these figures are controlled for all other variables.

function is nonseparable. The problem with a nonseparable damage function is that any policy which depends on assessing the damage caused by each pollution source is made much more expensive.

The separability of the damage function, however, is a distinct phenomenon from the separability of firm's cost functions or consumer's utility functions. To see this more clearly assume there are two upstream polluters which only affect a downstream firm. If one of the upstream firms is the only polluter in the first instance, then the downstream firm will suffer damage. If the cost function of the downstream firm is separable in pollution, the damage is in the nature of a fixed cost which affects only its profit position, and not its choice of output. If on the other hand its cost function is nonseparable, the damage is equal to the loss of revenue net of costs on the marginal adjustments it made in its output level (see Figure 4-12). Now imagine the calculation of losses to the downstream firm if the other upstream firm is the only source of pollution. Again there is the case of separability and nonseparability to consider in the downstream firm.

If, now, both upstream firms pollute, it may be that the profit position of the downstream firm is altered by more than the sum of effects imposed by the separate upstream firms. However, the downstream firm may still maximize profits at the same output as when none, or one, of the upstream firms was polluting. In this case the pollution damage function is nonseparable, but the downstream firm's cost function is separable. It may be that the damage to the profit position of the downstream firm is just equal to the sum of the damage caused by the individual upstream firm's pollution while its output decision is

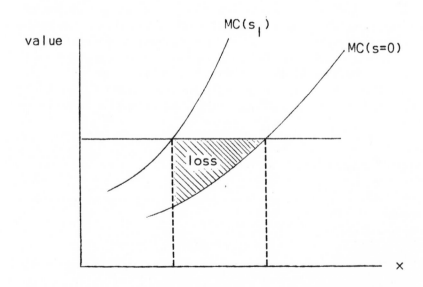

Figure 4-12.

unaltered. In this case both the damage function and the downstream firm's cost function are separable.

Moreover, in the case where the downstream firm's cost function is non-separable, the damage (net revenue lost on marginal output adjustments) may be equal to or greater than the sum of the damage due to the individual upstream sources alone. This occurs, respectively, if the damage function is separable or nonseparable.

Market Structure

The interesting fact about market structure in the pollution and congestion case is that, unlike the nuisance case, market structure does not matter. The optimum for a given problem shed is a quantity which minimizes the sum of damage costs and abatement costs. This minimum cost optimum does not require that the demanders of the pollution or congestion be competitors in their output markets.[10] While it is true that the precise optimum value may be different were the demanders competitors instead of monopolists, the quality of the optimal solution is the same. Namely, it represents a quantity of polluting activity which balances at the margin gains and losses from further reductions in the activity so that the social cost of the pollution is minimized.

But this appears to be a second best problem of which it is true that if any one of the Pareto conditions cannot be met, it is in general not desirable to fulfill the remaining ones.[11] Perhaps it is not desirable to achieve optimality in a problem shed as long as there is monopoly (or any other non-optimality) in the system. This is not the case here, however. If there is monopoly in the system, its price is greater than marginal cost leaving some unexploited consumer surplus. But whether optimality is achieved in a problem shed or not, that surplus will go unexploited. If the monopoly is institutionalized so that there is no effective way to deal with it, then that market structure is a binding constraint in the welfare maximum problem. As Morrison says, ". . . once an artificial restriction is accepted as binding there is no reason to distinguish it from natural restrictions."[12] It is true that as long as there is monopoly in the system, the optimality conditions in the problem shed differ in detail from ones that would obtain in an all competitive system. In general it is not desirable to achieve the conditions necessary for a first best optimum in the problem shed when there is a noncompetitive market structure. But it is desirable to achieve the second best optimality conditions in the problem shed, which are second best only with respect to the ideal market structure, but are "first best" with regard to the given market structure. "Since both the optimum and the second best are positions of constrained maxima, the differences between the two are not qualitative differences, but rather differences in detail."[13]

The conclusion is that market structure may be taken as given an attempting to achieve optimality in a problem shed. Whether or not the polluter is a monopolist, or anything else for that matter, is irrelevant in decisions with regard to the optimal use in a problem shed. The fact that market structure may be taken as given for policy regarding pollution does not mean that existing market structure must be accepted as permanent. It means that for purposes of pollution policy (optimal use of a problem shed), market structure may be ignored. It is equally true that for the pursuit of policy regarding market structure (antitrust policy), the use of relevant problem sheds may be ignored. This contrasts with the nuisance model where it is impossible to ignore market structure in determining an optimal amount of the nuisance-causing output.

5 Policy to Achieve Optimality: Nuisance

Introduction

The nuisance case of nonmarket interdependence is the case where abatement possibilities are foreclosed. It is the output of firms or the consumption of consumers that directly imposes costs on another economic agent. Having explored the nature of an optimal allocation when a nuisance exists in Chapter 3, we are now in a position to ask about policy. Since it was demonstrated that private maximizing behavior fails to achieve a Pareto optimal distribution of resources, the problem we face is what to do in order to secure a more nearly optimal allocation. The expected results of various alternatives are examined in this chapter.

Voluntary Action

In this section we explore potential actions which might be undertaken by parties to the technological externality of their own volition. These actions come under two main headings: bargaining and merger. Although any economic decision-making unit could in principle bargain if it were party to a technological externality, merger is a feasible alternative only for firms. In the second part of the chapter involuntary actions which are policy-induced are explored.

Bargaining and Liability

Bargaining takes place in a legal environment with established rules of liability. The outcome of the bargaining solution is examined with respect to rules of liability to see what might be expected under various circumstances. Liability is defined in the dictionary sense of being legally bound to make good a loss or damage that occurs and is not taken to mean, for example, a tax liability. The question of taxes is considered in a later section, so the question of the assignment of liability considered here does not involve the assignment of taxes. Rather, it means who is legally responsible, if anyone, for damage that occurs as the result of a nuisance. Two main cases are examined; one in which the damaging party is legally liable, and one in which the damaging party is not liable.

To begin examination of this question, consider a nuisance occurring between consumers unilaterally. Assume further that the nuisance-creating good is available to the two consumers in fixed supply and the question is one of allocating the consumption good optimally. We have seen in Chapter 3 that private maximizing behavior in free markets results in misallocation. Some writers, the most influential of whom is Coase,[1] suggest that the market misallocation can be dealt with by having the parties directly involved reach a negotiated settlement outside of the marketplace when negotiations involve negligible cost.

In Figure 5-1, the length of the abscissa represents the total amount of nuisance-creating x that is available for distribution between consumers A and B. B's consumption of x_1, however, causes a nuisance to A. The curve which is downward sloping from left to right represents A's marginal rate of substitution of x for some numeraire good, while the solid curve which is downward sloping from right to left represents B's marginal rate of substitution of x for some numeraire good. The dashed curve represents B's private rate of substitution plus the (negative) external rate of substitution, or the marginal social rate of substitution. The vertical distance between the private and social rates of substitution of B's consumption represents the external damage caused to A as a result of the nuisance. The socially optimal allocation at x^O is described analytically by

$$\text{MRS}_{21}^{a} = \text{MRS}_{21}^{b} + \text{MERS}_{21} \tag{5.1}$$

where MERS_{21}, the external rate of substitution or damage caused by B's consumption of x, is negative. We may write (5.1) as

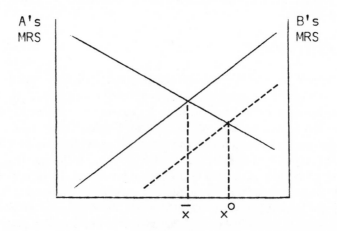

Figure 5-1.

$$\text{MRS}_{21}^{a} - \text{MERS}_{21} = \text{MRS}_{21}^{b}. \tag{5.2}$$

Since the quantity MERS_{21} is negative, we are actually adding something positive to MRS_{21}^{a} in the simple transposition of (5.1). The Equation (5.2) is represented graphically in Figure 5-2 with the dashed curve, A's net, representing the left-hand-side of (5.2). The arguments $\overline{x}$ and x^{o} are the same as in Figure 5-1 where the market equilibrium is given by $\overline{x}$ and the social optimum by x^{o}.

The argument is that at the market equilibrium, $\overline{x}$, A stands to gain an amount bc on the $\overline{x}$th unit plus an amount ab in damages not incurred if the $\overline{x}$th unit is allocated to A instead of B. A's total gain from reallocation of the $\overline{x}$th unit is ac. But B stands to lose an amount bc if the $\overline{x}$th unit is reallocated away from him to A, which means A stands to gain an amount ab more than B stands to lose. So, if A and B can bargain, B would give up the $\overline{x}$th unit to A for anything more than bc, while A would clearly be willing to compensate B anything greater than bc but less than ac. Some bargain will be struck on the $\overline{x}$th unit. Now, precisely the same argument can be made for the $(\overline{x}+1)$st unit. Some bargain will be struck on that unit in which A pays B something more than B's loss (ef) but something less than his gain (df). The limit to this type of bargaining is the social optimum, x^{o}. Beyond x^{o}, B stands to lose more than A would be willing to compensate, so no bargain can be struck. Starting from the

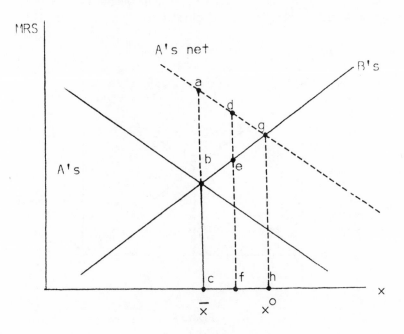

Figure 5-2.

suboptimal market equilibrium, however, a social optimum can be reached if bargaining is possible, and costless.

Coase goes one step further than this in the context of a slightly different model. Assume now, instead of x being a good in fixed supply which is allocated to A and B, that the good x is consumed by B and not A, but that there flows a nuisance from B's consumption of x to A. The difference between the two models is that in the present model the curve labeled A's in Figure 5-3 is simply the $MERS_{21}$ since there is no marginal rate of substitution, MRS_{21}^q, when A does not consume x directly. The activity x might be Coase's number of trains, or number of steers, or Buchanan and Stubblebine's height of the fence. The nuisance damage to A is respectively in these examples loss of crops by farmers along the tracks due to sparks, loss of trampled crops, or loss of view by a neighbor. The important feature of these models is that there is no abatement possibility.

Coase claims in what has come to be known as Coase's Theorem, that in this model, rules of liability do not matter.

COASE'S THEOREM. The allocation of resources will be the same in the case when there is no liability for damage as it is when the damaging business is liable for damage caused, when the parties can bargain costlessly.[2]

If there are no rules of liability against B, then he will pursue the activity to the point $\overline{x}$ where all benefits net of private costs are exhausted. But in this

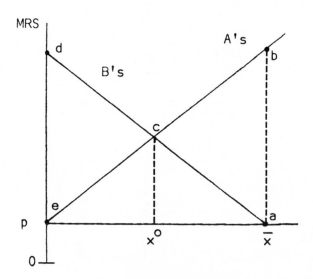

Figure 5-3.

situation, A will incur a loss of ab on the xth unit. But since B would accept anything greater than Oa to forego the $\overline{x}$th unit, some mutually profitable bargain can be struck on the $\overline{x}$th unit of consumption. Bargaining on all successive units up to x^Oth unit will occur in the same fashion with x^O being the limit after which no bargains can be struck.

Assume now that there is a rule which states that B may do nothing without incurring liability to A for damage. In this case the starting point is 0. But on the first unit of x, B would gain Od while A would lose almost nothing. Therefore, B would consume the first unit of x paying A an amount equal to his damage which is clearly less than de. B would continue to consume x up to x^O compensating A for his marginal damages. At x^O, B would gain cde, while A would lose (in damage) and gain (in compensation) an amount cex^O, leaving him indifferent. There is a clear Pareto gain to be had by moving from $x=0$ to $x=x^O$ under the new liability rule. Under either liability rule, the outcome is the same, according to Coase.

But this influential and oft-cited theorem is only right as a special case as shown by Mishan, Dolbear, and Burrows.[3] Their criticism is focused on the fact that, in general, there is an income effect which depends on the legal starting points and influences A's marginal rate of substitution. The best demonstration of this involves looking at what Dolbear calls the triangular Edgeworth box.[4]

To construct the triangular Edgeworth box, assume that consumer A lives by bread alone. B, on the other hand, cannot live by bread alone, but also needs heat. The heat causes smoke, and the smoke bothers A. Moreover, Dolbear is clearly dealing with the nuisance case when he states that, "[B] can only reduce the amount of smoke by decreasing heat."[5] Because heat and smoke go together, they may be measured on the same axis for appropriately chosen units. B's indifference curves, showing a preference field between heat and bread, have the standard shape if there is diminishing marginal utility to both consumption of bread and heat. B's MRS^b_{BH} is negative, but decreasing. But B's heat causes smoke for A, and since the smoke is nonabatable, we can think in essence of only one commodity, combustion, which is a good to B and a bad to A. Because combustion is bad to A, his MRS^a_{BH} is positive; A will accept more combustion only with more bread to remain equally well-off. Or, said another way, the nuisance is normal as defined by Diamond and Mirrlees. Because of diminishing marginal utility, the shape of A's indifference curves are as depicted in Figure 5-4, Plate a.

Using the optimality conditions for a public good (in this case combustion),

$$MRS^a_{BH} + MRS^b_{BH} = MRT_{BH}, \tag{5.3}$$

assume the marginal rate of transformation is a constant given by the slope of FF and transpose (5.3) to get

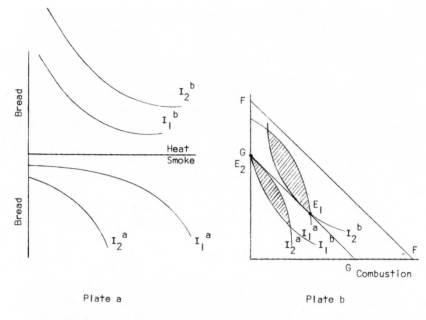

Plate a   Plate b

Figure 5-4. Source: F.T. Dolbear, "On the Theory of Optimum Externality," *American Economic Review* 57 (1967): pp. 92 and 95.

$$\mathrm{MRS}^{b}_{\mathrm{BH}} = \mathrm{MRT}_{\mathrm{BT}} - \mathrm{MRS}^{a}_{\mathrm{BH}}. \tag{5.4}$$

Now superimpose the lower quadrant of Plate a, Figure 5-4 onto the upper quadrant redrawing the indifference curves for A to display a slope equal to the right hand side of (5.4). The result is Plate b, Figure 5-4. Since the slope of B's indifference curves is equal to the left hand side of (5.4), tangencies between the two sets of curves represent Pareto optimal allocations, i.e., allocations for which (5.4) is met.

Now for Dolbear's demonstration. Given an initial endowment of resources with production possibilities given by *FF*, let the line *GG* represent an arbitrary but definite ethical distribution of income. B can choose the combination of heat and bread he wants along *GG*, while A gets the amount of smoke chosen by B and an amount of bread represented by the vertical distance between *GG* and *FF*. Corresponding to the liability rule which says that B can do anything he desires with impunity, there is an equilibrium point E_1 for B. At E_1, B has achieved the highest possible level of satisfaction (I_2^b), subject to his budget constraint (*GG*). But at E_1, A is on indifference level I_1^a and receives bread in the amount represented as the difference between *GG* and *FF*, and he also receives an externality from B in the form of smoke generated in the amount given by the abscissa component of E_1. But at E_1, notice that there is no tangency between I_2^b and I_1^a.[a] This means that the liability rule favoring B does not yield a Pareto optimal allocation. If A and B are free to bargain, the bargaining space is clearly the shaded area between I_1^a and I_2^b.

Now examine the allocation which obtains under the liability rule which says B may do nothing without A's consent which, in effect, makes B liable to A for smoke damage. In this case, there is an equilibrium starting point at E_2 where no smoke is generated and the bread is divided according to the distribution of income. But E_2 is not a Pareto optimum either,[b] as I_1^b and I_2^a are not tangent. In this case the bargaining space is represented by shaded area between I_2^a and I_1^b.

The contract curve of Pareto optimal allocations which represents the locus of tangencies satisfying (5.3) is given in Figure 5-5 by curve JJ'''. The solid portions of JJ''' are those portions that lie within the respective bargaining spaces. What the final allocation of bread and combustion will be is uncertain and depends upon the bargaining skills of the negotiating parties. But we can say that under the liability favoring B (starting point E_1), the final outcome will lie along JJ'. Under the alternative liability rule favoring A (starting point E_2), the final outcome will lie along $J''J'''$. In general, liability rules do affect the amount of a nuisance, contrary to Coase's Theorem.

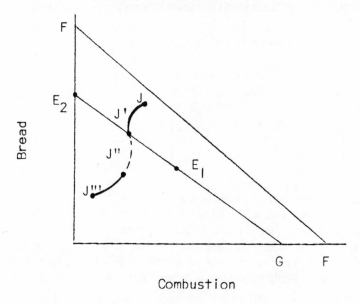

Figure 5-5.

---

[a]The slope of I_1^a is MRT $-$ MRS$_{BH}^a$, so that for any positive value of MRS^a the slope of I_1 is steeper (in a negative direction) than the MRT (the slope of GG).

[b]That is, unless I_1^b lies everywhere above I_2^a, in which case E_2 is a corner optimum.

The only time that the amount of combustion, and hence smoke, is invariant with respect to rules of liability, as Coase claims, is when the contract curve JJ''' is a vertical line. This is true when all of A's indifference curves have the same slope at each given level of combustion. Stated another way, JJ''' is a vertical line when, and only when, A's marginal rate of substitution is independent of the quantity of bread, i.e., when there is no income effect on A's preference for smoke. Although Dolbear provides a neat and imaginative demonstration of this fact, Mishan[6] before him and Burrows after him recognize the same phenomenon.[7] Liability rules do affect allocation except in the special case when there are no income effects. We now have the rationale to formulate the

MISHAN-DOLBEAR-BURROWS THEOREM. The allocation of resources is not independent of the rule of liability governing a situation where the parties can costlessly bargain, except in the case where there are no income effects.

But even if there are no income effects, Coase's result must be qualified. Marchand and Russell[8] show that the issue of separability enters into the question of whether or not liability matters. They examine the case of a unilateral nuisance in production. Suppose firm 1's output enters in an adverse way into the cost (and hence profit) function of firm 2. They assume

$$C^1 = C^1(x_1) \text{ and } C^2 = C^2(x_1, x_2) \text{ where} \tag{5.5}$$

$$C_2^2 > 0, \ C_1^1 > 0, \ C_1^2 > 0,$$

$$C^1(0) = 0, \ C^2(x_1, 0) = 0,$$

$$\begin{vmatrix} C_{11}^1 + C_{11}^2 & C_{21}^2 \\ C_{12}^2 & C_{22}^2 \end{vmatrix} \neq 0 \text{ for } x_1, x_2 > 0,$$

where the subscripts indicate partial derivatives.

Their results are tabulated in Table 5-1, but the force of these results is that even ignoring income effects, Coase is right only if the cost and utility functions are separable. In the event that the cost or utility functions are nonseparable, allocation is dependent on the rule of liability.[9] This means that the Coase Theorem is substantially weakened in a way not recognized in the original formulation.

Calabresi attacks the Coase result on still another front. His criticism is that "The short of the matter is, liability rules do affect the amount of money people make (rents) in the short run; and in the long run, people will enter those activities where they make more money."[10] In the long-run the allocation of

Table 5-1

State	Separability	Nonseparability
Individual Maximization ($\hat{x}$)	$x_1 > x_1{}^*$	$x_1 > x_1{}^*$
	$x_2 = x_2{}^*$	$x_2 < x_2{}^*$
Liability on Firm 1 (x')	$x_1{}' = x_1{}^*$	$x_1{}' < x_1{}^*$
	$x_2{}' = x_2{}^*$	$x_2{}' > x_2{}^*$
No Liability on Firm 1 (x'')	$x_1{}'' = x_1{}^*$	$x_1{}'' > x_1{}^*$
	$x_2{}'' = x_2{}^*$	$x_2{}'' < x_2{}^*$

Let x^* be the optimal output (joint profit maximum)

Source: J.R. Marchand and K.P. Russell, "Externalities, Liability, Separability, and Resource Allocation," *American Economic Review* 63 (September 1973): 611-20.

resources is not invariant with respect to liability rules because of the long-run industry adjustment to rent differentials caused by liability rules. But then, Calabresi makes a concession on this point by saying that any long run "misallocation" due to an altered industry mix and output configuration as a result of changing the liability rule could be corrected by costless bargaining.[11] It appears at this point that the discussion loses focus of the difference between misallocations and nonunique optima. The fact that under one liability rule the distribution of rents is different, and hence the optimal industry and output mix is different than under some other rule is not to say that either resulting mix is a misallocation. With respect to differing liability rules, one might see differing optima due to long-run adjustment to differential rents even after costless bargaining is permitted to run its course. And that is exactly the point Calabresi set out to make originally.[12]

Is it still true that costless bargaining can improve the market allocation when a technological externality exists? The answer, of course, is yes; but this is true in any and all circumstances. That is, it is true by definition, which gives rise to what Calabresi calls a "Say's Law of Welfare Economics."[13] A nonoptimal allocation is defined to exist when

... there is available a possible reallocation in which all those who would lose from the reallocation could be fully compensated by those who would gain, and, at the end of the compensation process, there would still be some who would be better off than before.[14]

Or, stated another way, a nonoptimal allocation is defined to exist when costless bargaining can improve the situation. But which optimal solution is the result of the bargaining process depends on the separability of cost and utility functions, on the liability rules which create differential income effects and demand patterns, as well as differential distribution and hence long-run industry and output mix. Which liability rule should be chosen? Recognizing implicitly that liability rules do affect the optimal outcome in general, Zerbe indicates that "The rule that emerges is that liability should be imposed so as to obtain the pattern of adjustments that yields the greatest net social product."[15]

But there are still more issues concerning the bargaining solution to the nuisance problem. One problem concerns the nature of the bargaining process itself. The view of bargaining that must be maintained in order for it to result in optimality is that bargaining is a marginal process. If a bargain is struck on the last unit, then the next unit becomes the subject of negotiations. It is as if the two neighbors fighting over the height of a fence bargain inch by inch to a solution. But the bargaining limits, says Wellisz, are not set by what is to be gained or lost on the marginal unit. Rather, the bargaining limits are set by the total amount that is to be gained or lost by a discrete, finite move away from the status quo. In a bargain that is not reached as a marginal process, there is no guarantee that it will result in an optimal configuration.[16] Wellisz concedes that if a large number of externality producers compete for victims, competition insures that bargaining is restricted to the marginal unit. But the connection of this case with reality is tenuous; Wellisz says,

When, in the absence of zoning, a factory is erected, it is not usual for the factory owner to notify the nearby farmers about the amount of smoke which he intends to produce, so that bargaining, if any, takes place after the factory is built. It is still more unusual for a number of factories to compete not only for a site but for the privilege of smoking on nearby fields.[17]

Therefore, in most real world cases in which externality producers do not compete for victims, bargaining will likely not take place as a marginal process, and the resulting bargaining solution will likely not be optimal. It might be argued that if the negotiating parties consider all finite moves from the status quo, and if they reveal to each other their true bargaining limits, it will soon become clear that the best bargain will be the bargain reached at the optimum. But part of the process of good bargaining is not to reveal the exact bargaining limits. In this circumstance, there is a whole number of outputs on which mutually profitable bargains can be struck. The point is that there is nothing driving the bargainers to the optimum output in the absence of omniscience and if bargaining is a discrete process. This seems to be the same point made by Arrow in a different context. In the famous lighthouse example, if ships come into the range of the light one at a time so that exclusion is no problem, (the lighthouse owner need merely shut off the light) there are still no forces which would drive a shipowner and the lighthouse owner to an optimal result.[18]

A closely associated problem is the problem of threats. This is an outgrowth of the fact that good bargainers never disclose their true bargaining limits, and the realization that bargaining is rarely a marginal process. In Coase's train example, this means that the railroad might threaten five trains per day instead of bargaining with the farmers over the second (marginal) train. Wellisz[19] and Shoup[20] realize the threat potential, and Marchand and Russell[21] indicate that their results depend upon the fact that threatened and actual quantities are identical. Mumey shows that threatening always leads to a preferred position, and further that "... an economic incentive is demonstrated for the development of potential actions that can inflict harm on others, even if they concommitantly would inflict internal damages."[22] Moreover, there is what Mumey calls "malicious damage" which is defined to be costs imposed as a result of actions which yield no benefits (except the possibility of extorted money) to the perpetrator.[23] There are laws against this kind of extortion, as Zerbe[24] and Demsetz[25] point out, in addition to which threats can go either way.[26] Demsetz further relies on the notion of competing extortionists to offer to refrain from their activity for less and less, and ultimately a zero amount.[27]

But even within the framework of a legitimate nuisance case, and not the "malicious damage" of robbery, protection, blackmail, or illegal extortion, the fact that threats can go either way does little to insure that the bargainers are propelled to the optimum in a bargaining situation. Furthermore, the notion of extortionists competing themselves out of business is only tangentially relevant to the problem if threats can be carried out. If an armed robber threatens to shoot me unless I give him $100, the fact that there are a lot of other robbers around who would shoot me for less is irrelevant.

If we get away from legitimate nuisance into the area of malicious damage, we are even further away from guaranteeing optimality. Mumey argues that the issue of negotiating a limit to malicious damage is not merely one of distribution, as some have argued.

... If the threats of malicious damage were allowed to become instant income generators, the collection of bribes could become the principle modus of income distribution. Since the allocation of resources in a voluntary economic system is inextricably bound to that system's pattern of reward dispensation, one finds it hard to believe that [allocation] could emerge unscathed if an essentially coercive income distribution system were to become pervasive.[28]

The defense of Coase against the threat issue by saying that threats cost money and therefore violate the costless bargaining assumption[29] seems shallow at best. Whether or not threats are costless, their use in general leads to a suboptimal bargain being struck. Even if a cost is incurred to threaten, a bargainer can negotiate a suboptimum bargain which is preferred over the bargain reached without the threat, as Mumey shows. Thus, the incentive exists to make threats, which, once made, greatly reduce the chance that a bargain will be optimal.

The final consideration in bargaining as a policy prescription in the case where nonabatable externalities intervene with market allocation is that precisely because there are likely to be costs in reaching and enforcing a bargain, any gain due to the reallocation must be offset by the cost of obtaining it to determine if the move is worthwhile.[30] Moreover, the free-rider problem is a big consideration in the determination of the group will.[31] In order to ascertain of a given group what it would be willing to pay to get rid of ten, twenty, . . . , one-hundred percent of the nuisance, the group organization must ask each individual in the group what it is worth to him. Of course, if the organization is to bargain with the externality-creating party, it must not only know the worth of reducing the nuisance, but it must be able to produce that worth as a bribe in exchange for a reduction. So, the group organization must ask each individual what a reduction in the nuisance is worth, and, having determined that, it must ask the individuals to hand over whatever it is worth. But each individual making a decision will be induced to understate the worth to him of any given reduction on the premise that if the group secures a nuisance reduction in the negotiations, he will benefit anyhow, regardless of what he pays, and if no allocation is secured, he will be no worse off than his present position. In short, he is a free-rider on the backs of the rest of the group. But if everyone behaves this way, no bargain will be secured at all.

Conclusion. The definition of a nonoptimal allocation of resources is that costless bargaining can leave at least one individual better off in a net sense (i.e., the social product is smaller than its maximum if there is a nonoptimal allocation). A situation characterized by a nuisance results in a nonoptimal allocation of resources when individuals behave as private maximizers. Therefore, it is natural that costless bargaining could result in an optimal allocation. But which of the nonunique optima is reached depends in general upon the liability rules governing the bargaining situation when there are income effects. Moreover, even in the absence of income effects, liability rules affect the optimal outcome if the cost or utility functions involved are nonseparable. Since liability assignment affects the transfer of rents and consumer surplus, long-run demand patterns, industry, and output mix are affected by a liability assignment. In general the world will look different under one set of liability rules than under another. The possibility of threats makes a bargaining solution less likely to be a socially optimal bargain. And if bargaining is not a marginal process, the likelihood of an optimal bargain being reached is also reduced. Finally, bargaining is likely to involve transaction costs, especially in the large numbers case, and therefore the gains from reallocation must be weighed against the costs.

Merger

As noted earlier, merger is a viable solution to the nuisance problem only if the parties involved are firms. It is hard to imagine what merger means in the case of consumers' or consumers' and producers' interdependence. Davis and Whinston make the case that powerful market forces exist which may lead to merger whenever two firms are linked by a technological externality.[32] Their contention is that, ". . . a tendency toward such mergers is consistent with, if not implied by, the idea of profit maximization."[33] This is true because the profits accruing under joint management are greater than the sum of the profits of the individual units under separate management. While the proposition that profits are greater under joint maximization than they are under individual maximization is true, from the stockholders' point of view it may be that some other action would yield even greater profits. For example, one firm might seek and obtain a permanent court injunction against the other firm.

Consider the situation of two firms that are a nuisance to each other, where the cost functions (for computational ease) are separable. Let their respective profit functions be

$$\pi_1 (x_1, x_2) = 5x_1 - x_1^2 - x_2 \qquad (5.6)$$

$$\pi_2 (x_1, x_2) = 5x_2 - 0.5x_2^2 - x_1.$$

Individual profits, social product, and profits per share are computed and tabulated in Table 5-2 under four alternative states of the world: (A) firms maximize profits individually, (B) firms merge and maximize joint profits, (C) firm 1 prevails in a suit against firm 2, and (D) firm 2 prevails in a suit against firm 1. In case B, note that profits per share is actually 6. But the shares of the new, merged firm will be distributed in such a way that the owners of both firms earn more than before. For example, if the two shares are distributed so that owners of the former first firm get 0.25 shares while the owners of the former firm 2 get 1.75 shares, then the return to the former first firm is 1.50 and the return to the former second firm is 10.50. These states are ranked according to the Pareto criterion of maximum social product and the individual criterion of profits per share. Of course, joint maximization afforded by merger is the socially preferred outcome. But, if a firm had its choice between merger and winning a permanent court injunction against the other firm, it would clearly choose the latter state.

When a firm initiates a suit, however, the probability is not unity that it will prevail. Moreover, since there are mutual nuisances, a suing firm might expect a countersuit which carries with it a finite probability that the firm will be prevailed against. If we assume that the sued firm files a countersuit in all cases, then when would one firm sue the other? There are three possible outcomes: (1) the suite is won (and the countersuit is lost), (2) both the suit and the countersuit are lost (and firms return to the initial state of individual maximization), and (3) the suit is lost but the countersuit is won. The firm must make a probability estimate of the likelihood of these outcomes if it is to determine an

Table 5-2

State	Individual Profits	Social Product	Profits per Share[1]
A. Individual Maximization (Initial State)	$\pi_1{}^A = 1.25$ $\pi_2{}^A = 10.00$	$\Sigma \pi_i{}^A =$ 11.25	$P_1{}^A = 1.25$ $P_2{}^A = 10.00$
B. Joint Maximization	$\pi_1{}^B = 4$ (as π-center) $\pi_2{}^B = 8$ (as π-center)	$\Sigma \pi_i{}^B =$ 12.00	$P_1{}^B = 1.50^2$ $P_2{}^B = 10.50^2$
C. Firm 1 Prevails Against Firm 2	$\pi_1{}^C = 6.25$ $\pi_2{}^C = 0$	$\Sigma \pi_i{}^C - \pi_2{}^A =$ -3.75	$P_1{}^C = 6.25$ $P_2{}^C = 0$
D. Firm 2 Prevails Against Firm 1	$\pi_1{}^D = 0$ $\pi_2{}^D = 12.50$	$\Sigma \pi_i{}^D - \pi_1{}^A$ 11.25	$P_1{}^D = 0$ $P_2{}^D = 12.50$

I. Social Ordering of States (Pareto Criterion)

$\boxed{B} > A = D > C$

II. Individual Ordering of States (Profits per Share Criterion) Unadjusted for Risk

Firm 1: $C > \boxed{B} > A > D$

Firm 2: $D > \boxed{B} > A > C$

[1] Each firm is assumed to have one share outstanding.
[2] Some distribution of the two shares can be found that yields greater return to both sets of shareholders.

expected value of a suit. Suppose that ρ_w^i is the ith firm's subjective probability estimate that it will prevail against the other firm, δ_L^i is the ith firm's estimate that both its suit and the countersuit will be lost. Then E_1 and E_2 given by

$$E_1 \text{ (suit)} = \rho_w^1 \pi_1^C + \delta_L^1 \pi_1^A + (1 - \rho_w^1 - \delta_L^1) \pi_1^D - C_1 \qquad (5.7)$$

$$E_2 \text{ (suit)} = \rho_w^2 \pi_2^C + \delta_L^2 \pi_2^A + (1 - \rho_w^2 - \delta_L^2) \pi_2^C - C_2$$

are expected values of a suit to firm 1 and firm 2 respectively if the sued firm files a countersuit, and where C_i is the legal expense to the ith firm. The question is what course of action will the firms follow? This depends, of course, on the expected value of a suit, its variance, the risk preference of the firms, and the per share profits in the alternative states of the world, but the results are summarized in Figure 5-6 for a risk averse firm. It should be evident from those results that it is not clear from the stockholder's point of view that merger is the outcome which is always preferred. When suit is brought, moreover, judges often do not decide a case on the basis of maximizing the social product, but rather on the basis of precedent, "who got there first," or other legal criteria, as the cases illustrate to which Coase alludes.[34] These legal criteria are irrelevant in terms of the optimal outcome, but there is no reason to suspect that the suit (and countersuit) will result in anything other than an injunction against one or the other of the firms, or a decision dismissing both the suit and the countersuit (because the nuisance is not "unreasonable" or there is "insufficient evidence").

The thrust of this discussion is that if the firms operate in a legal environment where injunctive relief is not provided for, then it is clear that there is a strong incentive to voluntarily merge. But if the legal environment provides for involuntary transfers (injunctive relief), the merger motive is substantially weakened.

Involuntary Actions

In this section we explore policy-induced actions designed to improve the allocation of resources when a nuisance exists. The policies explored here

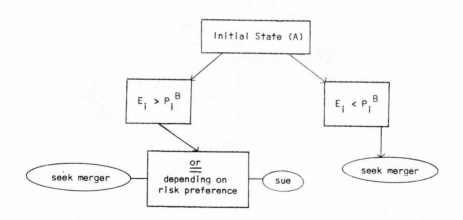

Figure 5-6.

are tax-subsidy approaches and alternatives which favor prohibitions and
directives.

The Tax-Subsidy Approach

The tax-subsidy approach is discussed in terms of Model 1 in Chapter 3, the
model of nuisance in consumption, but the discussion generalizes to the other
models as well. The tax-subsidy approach has been prescribed as a policy tool in
dealing with externalities at least since the time of Pigou,[35] and therefore the
application of taxes and subsidies to allocation problems involving externalities
of either the technological or the pecuniary variety has been termed the Pigovian
approach. The context of Model 1 in which B's consumption of x_1 causes a
nuisance to A, the Pigovian approach in its simplest form is depicted in
Figure 5-7. The solid, downward sloping curve represents B's marginal evaluation
of his own consumption of x_1, and it is the function along which he will respond
if left alone in a market. Because of the nuisance that B's consumption causes to
A, the social benefit of B's consumption is something less than his private

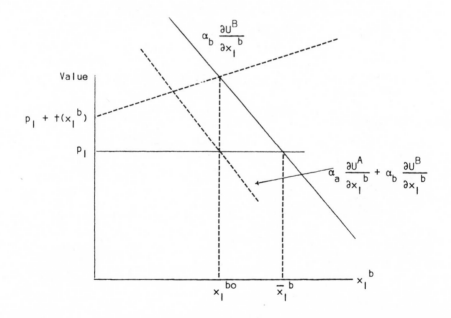

Figure 5-7.

evaluation of it, and this social marginal benefit function is depicted as the dotted, downward sloping curve. It is clear from the discussion in Chapter 3 that the socially optimal amount for B to consume is x_1^{bo} whereas the amount he will consume if left alone in a market in which the unit price of p_1 is charged is $\overline{x}_1^b$.

The problem is to secure an optimal allocatin of x_1^{bo}. The Pigovian approach to this problem is to place a unit tax on B's consumption of x_1 equal in amount to the marginal external disutility caused by the nuisance. If the tax is placed on B's consumption and B is left to respond in a maximizing way in the context of a market, he will be induced to restrict his consumption to the socially optimal level of x_1^{bo} where the marginal benefit of the last unit consumed is equal to its net marginal cost, $p_1 + t$.

This approach has come under attack not only from those who claim the cost of administering a tax subsidy would outweigh the gains from an optimal allocation, but also from those who claim that even if the cost of administration were zero, there are theoretical problems with the approach. The most devastating attack is leveled by Davis and Whinston, who claim that in the bilateral, nonseparable case, it is impossible to derive Pigovian taxes.[36] The crux of their argument hinges on a game-theoretic interpretation of the decision-making process. In the case of a separable, bilateral nuisance, assume that the two consumers have a choice of two consumption "strategies." Let a payoff matrix $\lVert [\, S_{ij}^A,\, S_{ij}^B]\, \rVert$ have typical elements

$$(S_{ij}^A, S_{ij}^B) \tag{5.8}$$

where the first element is the consumer surplus available to A when A chooses strategy i and B chooses strategy j, and the second element has a similar interpretation except for B.[c] Davis and Whinston's point is made in terms of the following example of a payoff matrix:

$$(5.9)$$

		B's Consumption	
		1	2
A's Consumption	1	(0.9, 0.9)	(0,1)
	2	(1,0)	(0.1, 0.1)

In terms of this matrix, A's choice for consumption of the nuisance causing good is represented by a row choice, while B's choice is equivalent of a column choice.

Notice that A's problem is to compare the first element in each of the pairs across rows. So, for example, should B choose strategy 1, A's surplus is 0.9 with

[c]Davis and Whinston's discussion is in terms of two mutually annoying firms, with the payoff matrix representing profit instead of consumer surplus. For consistency, the case of consumption is developed here instead.

strategy 1 and 1 with strategy 2. A chooses 2 regardless of B's choice of output as can be seen by making the same comparison when B chooses strategy 2. This is no accident of construction, but rather it occurs because of the fact that both of the utility functions are separable. A's marginal decision is independent of B's actions.

The same argument holds for B. Regardless of which strategy A chooses, B clearly prefers column 2. The fact of separability implies the game theoretic notion of dominance. Thus the private solution is (A2, B2). But a joint maximum of surplus occurs at (A1, B1). Davis and Whinston admit the possibility that a Pigovian solution which would induce A and B to change their consumption to the social optimum is possible in the separable case.

The problem according to them is with the nonseparable case. If both utility functions are nonseparable, then the marginal decision of each regarding his own consumption is influenced by the consumption of the other. In this case there is no game-theoretic property of dominance in the choice of strategies. A's strategy choice affects B's choice, and B's choice in turn affects A's. What the private equilibrium will be is difficult to say a priori, moreover, the ugly possibility of nonexistence of a private equilibrium emerges.[37] In either case, their argument is that it is impossible to construct Pigovian taxes to internalize the externality when utility functions are mutually nonseparable.

A second line of attack comes from Buchanan,[38] who indicates that a unit tax on a nuisance might actually decrease welfare if the externality is a nuisance in production and if the producer is anything but a perfect competitor. This line of criticism is well taken, and its implication for policy with a view toward an optimal allocation of resources is explored in Chapter 3 under the section entitled, "Market Structure and Nuisance."

A third line of criticism is brought by Buchanan and Stubblebine[39] and Turvey.[40] Their argument is that in the case parties to the nuisance can bargain, the placement of a tax without the distribution of the proceeds to the damaged party will result in a suboptimal allocation. Their argument is as follows: In private equilibrium, B will exhaust all benefits to him net of price; this occurs at $\bar{x}_1^b$. But because of the nuisance to A the optimal consumption of B is at x_1^{bo} where the social benefits represented by the dotted line are just exhausted net of price. But if B is charged an amount equal to the external marginal disutility of A, his *net* benefit function is just the social benefit function (the dotted line). Moreover, charging the tax does not eliminate the damages, so that A's marginal damage function is still valid in reflecting the harm B causes to A. Therefore, if A and B can negotiate, A would induce B to cut back to $\hat{x}_1^b$ by offering anything up to his marginal damages for reduction of successive marginal units. Since between x_1^{bo} and $\hat{x}_1^b$ this amount is greater than the amount B would derive from consuming the unit, a bargain will be struck on all units between x_1^{bo} and $\hat{x}_1^b$ The new equilibrium with the tax when bargaining is possible is $\hat{x}_1^b$, clearly a suboptimum.

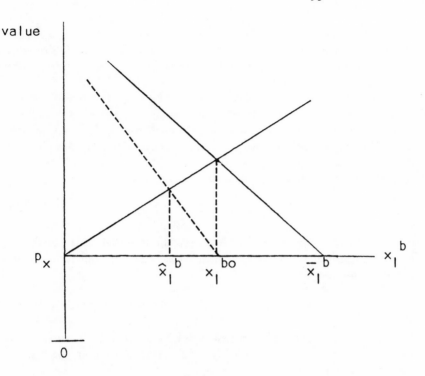

Figure 5-8.

On the other hand if the tax is distributed to A in the amount of his marginal damages so that he is exactly compensated, the net marginal damage function after compensation is paid is equal to zero. Now if B is taxed in the amount of the gross external damages, he will exhaust all of the net benefits (represented by the dotted line) net of price. This clearly occurs at x_1^{bo}. Since A has no net damages, there is no inducement to reach a bargain which would lead away from the optimum.

In response to this criticism, it may be noted that if bargaining can occur between A and B, then there is really no need to impose a tax on B. Let the two of them bargain to a solution, which under the most favorable circumstances (bargaining occurs at the margin and no threats) will result in an optimum. The only excuse for a tax from the standpoint of policy is to induce optimal behavior when optimal decisions are not forthcoming voluntarily. As a theoretical curiosum, however, it is true that if marginal bargaining can occur, and if a tax in the amount of marginal damages is imposed, then the tax must be distributed to the damaged party in the amount of his damages or else a suboptimal bargaining equilibrium will result. In no other case need the tax proceeds be distributed to the damaged party for optimality to occur. In fact,

Dolbear[41] proves that in the case of a nuisance in consumption, it is not possible to devise a unit tax that will at the same time raise sufficient revenue to exactly compensate the damaged party and induce a Pareto optimal distribution of resources.

In response to Davis and Whinston's contention that it is impossible to devise a Pigovian tax in the case of bilateral, nonseparable externalities, Wellisz presents an existence proof that such taxes must indeed exist.[42] Rather than follow Wellisz's proof of existence, we can actually derive optimal Pigovian taxes for nuisance in consumption after Diamond.[43] If x^h is the hth consumer's consumption of x, then let his utility function be

$$U^h(x^a, x^b, \ldots, x^h, \ldots, x^n) + y^h \tag{5.10}$$

where y^h is the income available for expenditure on all other goods but x. Nuisance in consumption occurs when consumption by others causes h to experience disutility.

$$\partial U^h/\partial x^j < 0, \, h \neq j. \tag{5.11}$$

In order to derive optimal taxes, first derive the demand functions for x^h as a function of net price, that is price plus tax. These come from the first-order conditions for a solution to the problem

$$\text{maximize: } U^h(x^a, x^b, \ldots, x^n) + y^h \tag{5.12}$$

$$\text{subject to: } (p + t)x^h + y^h = m^h$$

where t is the unit tax on x and m is the total income of h. The first order conditions, allowing for the possibility of a corner solution, are

$$\partial U^h/\partial x^h \leq p + t \text{ and} \tag{5.13}$$

$$x^h(\partial U^h/\partial x^h - p - t) = 0.$$

If the second order conditions are satisfied for the conditions (5.13) to represent a maximum instead of a minimum, then the demand functions exist and are decreasing in price:

$$x^h = x^h(p + t), \, \partial x^h/\partial(p + t) < 0. \tag{5.14}$$

In the nonseparable case where

$$\partial^2 U^h/\partial x^j \partial x^h \neq 0, \, j \neq h, \tag{5.15}$$

we can see from (5.13) that the marginal utility of own consumption is itself dependent on the values taken on by all other consumption variables. Thus the demand curve shifts with respect to changes in the consumption of others.

In order to derive the optimal Pigovian tax on x, form a welfare function and maximize it with respect to the tax. Let the welfare function be

$$W(t) = \sum_h U^h (x^a (p+t), x^b (p+t), \ldots, x^n (p+t)) \qquad (5.16)$$

$$+ \sum_h y^h .$$

where the demand functions used are the general equilibrium demand functions.[d] The use of general equilibrium demand functions accounts for the fact that the externality is nonseparable which means that everyone's consumption of x influences everyone else's demand. Now taking the derivative of W with respect to t and setting it to zero, we can solve for the optimal tax,

$$t^* = \frac{-\sum_i \sum_{h \neq i} (\partial U^h / \partial x^i)(\partial x^i / \partial (p+t))}{\sum_i (\partial x^i / \partial (p+t))} \qquad (5.17)$$

"The optimal surcharge is a weighted average of the externalities, $\sum_{h \neq i} \partial U^h / \partial x^i$, the weights being the price derivatives of demand ..."[44] In the simpler case of separability, the demand functions used in the welfare function (5.10) are invariant with respect to the consumption of others, and therefore they need not be specified as the general equilibrium demand functions.

The force of the Wellisz and Diamond papers is to show that Pigovian taxes do indeed exist in the nonseparable nuisance case, contrary to the assertion never actually proven by Davis and Whinston. No one has claimed that these taxes are easy or cheap to compute, but at least the economist is justified in maintaining the Pigovian approach in his kit of theoretically viable policy alternatives.

Baumol concurs in this result and in addition shows that the tax need not be distributed to the victims of the external nuisance in order for optimality to occur.[45] He also shows, as is implicit in Diamond's paper, that the victims

---

[d]The welfare function is the unweighted sum of the individual utilities. This type of welfare function is based on the ethic that each person's happiness is as important as the other's in the community welfare. Any other welfare function could be used without fundamentally altering the results so long as $\partial W / \partial U^h > 0$.

of the externality need not be taxed in order for optimality to occur.[e] Baumol's conclusion is that "... taken on its own grounds, the conclusions of the Pigovian tradition are, in fact, impeccable... Pigovian taxes (subsidies) upon the generator of the externality are all that is required [for optimality] ."[46]

Goetz and Buchanan introduce a special case in production which they claim will not yield to the Pigovian solution.[47] This is a case of reciprocal nuisance in production where all the parties to the nuisance are in the same industry, and all the cost curves are nonseparable. This case has no consumption counterpart because there is nothing that corresponds to the idea of an industry. The argument of Goetz and Buchanan proceeds by supposing that industry output is a constant, x^T, and that the costs of the representative firm in the industry are a function of its own output and the output of all other firms in the industry,

$$C^i = C^i(x_i, X_i), \quad \partial C^i/\partial X > 0 \tag{5.18}$$

where $x^T - x_i = X_j$. Fewer firms in the industry each producing a larger amount would cause X_i to be smaller than otherwise if the industry total, x^T, is fixed. But, fewer firms would mean a shift in the cost structure because of the nonseparability of cost functions.

The argument of Goetz and Buchanan is that the optimum will require fewer firms in the industry than the competitive solution in order to capture a reduction in external diseconomies. Thus, instead of the competitive cost structure, MC(X_O) and AC(X_O), the representative firm in the industry should realize the cost structure MC(X_1) and AC(X_1) in optimality. A unit tax on x would simply cause the suboptimal cost structure to shift upward and would permit the suboptimally large number of firms to continue to exist. There is no unit tax that would cause the correct optimal adjustment which is to have fewer, larger firms in the industry producing at x_i^O.[f] Thus, for any representative firm that does not get squeezed out of the industry in a move toward optimality, x_i^O represents the optimal output with fewer firms and a constant industry output.

There are several observations that need to be made with regard to this argument. First, with constant demand, the only way industry output can remain constant as the number of firms changes is if industry supply remains constant. The only way that industry supply can remain constant as the number of firms changes is for the nonseparability of the cost functions to assume a very special form, namely that the marginal cost functions of remaining firms shift exactly so as to compensate for the loss of a firm. Moreover, for industry output to remain constant regardless of the number of firms in the industry, the compensating shift of the remaining firm's marginal cost functions must be

---

[e]This proposition is more thoroughly treated in the charges-payments treatment in Chapter 6. Some writers claim that in the pollution (congestion) case the victim and the perpetrator of the externality need to be taxed for optimality.

[f]The authors everywhere use the optimality condition AC = MC.

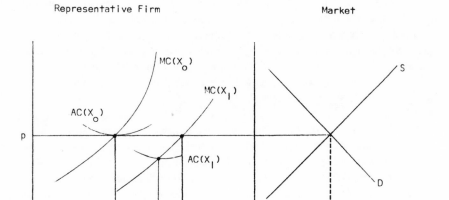

Figure 5-9.

ever-changing to account for the loss of firms of successively larger amounts of output. This is a very special assumption which we may expect in general not to hold.

Second, when the nonseparability is of precisely the form that would leave industry output unchanged as the number of firms changes, it is impossible for a profit maximizing firm to produce at the new $MC(X_1) = AC(X_1)$ equilibrium (with fewer firms) as Goetz and Buchanan indicate. The reason is that with industry supply and demand remaining unchanged, industry price will also remain unchanged. Therefore, a profit maximizing firm that remains in the industry will produce at $\hat{x}_i$ where price equals the new marginal cost. Because the existing firms will then earn excess profits ($p > AC(X_1)$) new resources will be drawn into the industry, causing a shift in the industry supply which will exert a downward pressure on price and cause industry output to expand. This will cause second order adjustments in the cost structure of firms as the industry output external to the firm (X_i) expands. The upshot of this is that no $AC = MC(=p)$ equilibrium can exist in the industry except the competitive one, unless industry output is allowed to adjust. Therefore, conclusions that depend on a constant industry output must be seen for what they are, namely, conclusions that apply in the most limiting of circumstances.

But if it is granted that industry output be permitted to adjust, might it still be the case that a unit tax is not sufficient to induce an optimal adjustment? Consider the fact that if the industry is a monopoly, then no external costs are incurred. Unit costs are lowest and production is economically most efficient when the industry is a monopoly. The industry is analogous for public policy purposes to other decreasing cost industries. Any given output is most cheaply supplied by having fewer and larger production units. Indeed, production is cheapest with only one producer. It appears that a unit tax is the wrong policy prescription in this case, just as it is in the case of other decreasing cost industries. A unit tax on the output of a competitive telephone industry would not secure the scale economies of production that are present. By the same token, a unit tax on a competitive Goetz-Buchanan industry would not cause the existing external scale economies of production to be captured. Policy should be aimed at encouraging monopoly, but as with other decreasing cost monopolies, care must be taken to prevent the gross abuses of monopoly power.

Prohibitions and Directives

Another class of policy actions designed to cause an involuntary adjustment consists of using administrative dicta sanctioned by legal authority. A prohibition is defined to be the legal authority to forbid the nuisance altogether. A directive is the legal authority either to require certain specific actions or forbid actions in excess of some standard.[48] For example, in the case of Buchanan and Stubblebine's fence, which causes utility not only to the fence builder but also to the neighbor, a prohibition would outlaw fences entirely. A directive might be in the form of a requirement that all fences be slat fences with 4-3/4 inch spaces between the slats. A directive forbidding actions in excess of a standard might take the form of forbidding fences in excess of 5-1/2 feet. The difference between the three is the degree of choice they permit. A prohibition permits no choice; the nuisance is simply outlawed. A directive requiring a certain specific action permits a bit more choice. While it may be thought of as a prohibition (prohibition, that is, of everything except the required action), there is the choice to meet the directive or not to engage in the activity at all. It is not an absolute prohibition. A directive forbidding actions in excess of some standard permits even more choice. The choice not to engage in the activity can be made, or the choice of many alternative actions within the limitation of the standard can be made.

The question is when do these administrative solutions maximize the social product? In the case of Model 1 in Chapter 3 dealing with consumption, a prohibition would give a correct result if the marginal social benefit lay everywhere below price. If the social marginal evaluation curve of B's consumption of x_1 is everywhere less than the scarcity value of x_1, then the optimality conditions of that model call for x_1^b to be zero. One way this corner solution can

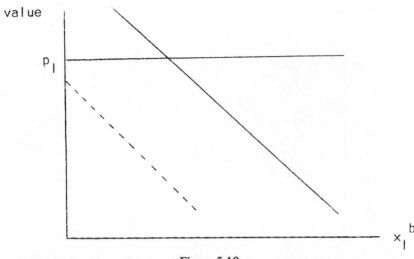

Figure 5-10.

be achieved is through a prohibition of B's consumption of x_1. In this case the prohibition would yield the socially desirable outcome.

There is another case when prohibition of a type might yield the socially desirable solution and that is the case of separate facilities.[49] Consider the case of tobacco smoking on an airliner at 39,000 feet altitude. Tobacco smokers will smoke until the benefits to them net of price are exhausted at $\overline{x}$.[g] But to nonsmokers, the amount of smoking $\overline{x}$ causes damages in the amount represented by the area under $D(x)$ to $\overline{x}$. Clearly, too much smoking takes place relative to the "optimal" amount x^O. At x^O the social product (benefits minus damages) is as big as it can be and is represented by the shaded area in Figure 5-11. What would happen, however, if smoking were prohibited in some of the seats in a contiguous portion of the plane? Then smokers and nonsmokers would segregate with the result that smokers could smoke as much as they want while nonsmokers would suffer no damage. The social product would increase to the total area under $B(x)$ to $\overline{x}$ which is a clear Paretian improvement. The same argument may hold for a number of cases. Beach areas might be divided between the surfing interests and the swimming and skin diving interests. Or the mutually antagonistic interests of sailors and speedboaters might optimally be resolved by prohibition, as they are on Lexington Reservoir in Los Gatos, California, by prohibiting speedboats on odd-numbered days of the month. The creation of separate facilities by prohibition, one might imagine, increases the social product

[g]Tobacco smoking is considered a nuisance (and not pollution) in our schema because no abatement is possible. The only adjustment possible is in the quantity of smoking. Language seems a little strained at this point, but hopefully the meaning of the language is clear.

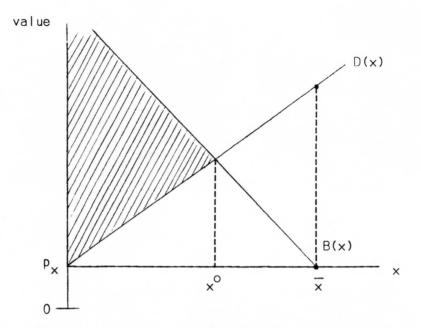

Figure 5-11.

of the water resource above what it would be if an attempt were made to secure an "optimal" mix of sail and power boats every day of the month. This idea is not new, as it is the basis for our zoning of urban property for certain uses. Yet it is often overlooked as a cheap and administratively easy policy alternative in dealing with nuisance problems.

Is it possible that all or most cases of nuisance can be resolved by providing separate facilities? If fence-haters can choose to live where fences are prohibited, there are no social costs and the net social product is the striped area in Figure 5-12. But a residential location is not only, or even primarily, chosen because fences are permitted or outlawed. Thus the social benefits of building fences is larger (at each height x) when people are allowed to live where they want to without regard to a fence zoning law. But then fence-builders and fence-haters will live in proximity to one another, and the fence-builder will cause the fence-hater some disutility. But it is possible, and in this case very probable, that the social product will be bigger by allowing people to live where they wish without regard to a zoning law but attempt to secure the optimal height fence, x^o, through other means. Separate facilities cannot work in every case.

When is it the case that a directive requiring a certain action is correct from the Paretian standpoint? Only when the required action is one which causes the

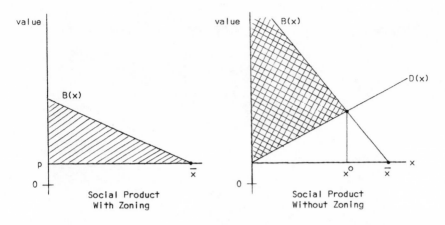

Figure 5-12.

first order conditions necessary for a Pareto optimum to be met. The requirement must be correct in both the qualitative and quantitative dimensions. For example, suppose the city fathers decide to require that all fences of whatever height be built as slat fences with 4-3/4 inches between the slats on the basis that such a fence serves the purpose of demarcating property and of keeping large domestic animals fenced, while at the same time not totally blocking the view of people on the other side. But if the fence-builder and fence-hater are like Buchanan and Stubblebine's, then the builder desires privacy, which is solely a function of the height of a solid fence, while the hater desires privacy up to a point but beyond that point prefers the view to more privacy. In this case, if the privacy-lover chooses to build a slat fence of any height, the privacy benefits to him are very much less than if he built a lower, solid fence. Moreover, the view is still partially damaged by a high slat fence over what it would be with a lower, solid fence. It appears that the action required by the directive is the wrong one if the criterion is to maximize the net social product. Other conceivable actions would yield a bigger product net of costs.

Legal authority to set and enforce a standard maximizes the social product if the standard that is chosen is exactly the quantity that causes the first order conditions necessary for a maximum to be satisfied. For example, in Buchanan and Stubblebine's fence case, a directive setting the fence height standard at 5-1/2 feet maximizes the social product only if 5-1/2 feet is the optimal height. Normally, however, the authority setting the standard does not know the marginal benefit and damages functions, B and D. But if it did, it will have noted that the cost-benefit ratio is smallest at the optimum. And in fact, it will have noted that the standard, whatever it is, is a function of the cost-benefit ratio.[50] Thus, in the absence of knowledge of D and B, it is true that the ". . . setting of

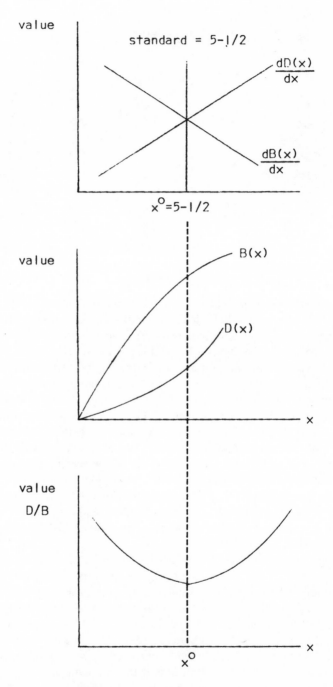

Figure 5-13.

a [standard] always involves a value judgment [valuation]; and this value judgment [valuation] always has the form of a cost-benefit ratio. *To set a [standard] is to impute a cost-benefit ratio* (italics in the original)."[51] This is true in the case of speed limits, fence heights or any other standard that is set to control externalities.

Information Requirements and Administration

The question now is what information is required for the various policy alternatives. In the case of bargaining and merger, nothing really needs to be known for policy purposes, because these are voluntary actions taken by the parties to the externality of their own volition. However, as noted before, if bargaining or merger do not take place in a legal vacuum, there are important informational requirements if a solution is to be evaluated according to a Pareto criterion.

For example, if firms can seek injunctive relief from other firms,[h] we would expect to see more injunctive relief sought and less merger than would be the case if injunctive relief were not a legal option. To the extent that firms prevail in seeking an injunction against other firms, the question must be asked whether the social product could be any larger by any other possible outcome. For example, imagine two firms which merge in a legal environment which does not provide for injunctive relief. If the new, joint venture shuts down one of the "divisions" in order to maximize joint profits, then we can conclude that the injunctive solution is indeed the optimal one if the legal environment provides such a possibility. But to be able to know whether or not one of the "divisions" would be shut down if a merger takes place requires knowledge of the cost and revenue functions of both firms so that the joint maximum solution may be calculated. So to the extent that injunctive relief is sought and granted in cases where shutting down one of the "divisions" is not a joint maximum, the legal solution does not afford an optimal one. Moreover, since all future firms are subject to the statutes, in order to be certain that a statute providing for injunctive relief were optimal, cost and revenue functions of all future firms that would seek such relief have to be known. Clearly this is impossible, so it cannot be said with certainty that in the case of firms, a statute providing for injunctive relief from what we have defined as a nuisance is, in fact, optimal.

In order to know whether or not a bargaining solution is optimal, it helps to

[h]By injunctive relief is meant a court order forbidding the production of the output causing the nuisance altogether. While from a legal point of view an injunction could stop production of the nuisance-causing output above a certain level, it is assumed the firm seeking the injunction experiences a marginal externality starting with the first unit of output. Therefore from an economic point of view, the damaged firm is likely to ask for the other firm to be enjoined from producing any output by which it is damaged.

know how many parties there are to the bargain. If, for example, the bargain is struck between two parties, we can be certain that there is nothing which compels the solution to be optimal. If bargaining takes place among a large number of parties so that there is in effect a market for the external effect, we can expect an optimal outcome in the most favorable of circumstances.

In order to compute optimal taxes on commodities causing a nuisance, it is sufficient to know utility functions of all individuals (production functions for firms), and general equilibrium demand (cost) functions.[52] In short we need all of the information required to solve the centralized allocation problem.[53] If this is the case,

... one may ponder whether there is any advantage in [the Wellisz-Diamond] proposal [for taxes] over having the authority dictate that the firms produce [or the consumers consume] the appropriate amount.[54]

After all, the main advantage of a tax scheme is that it permits decentralized decision-making and economizes on information. But if the information requirements to compute an optimal tax are as demanding as the requirements to solve the centralized allocation problem, the informational advantage of the tax scheme is lost.

Davis and Whinston[55] and Starrett[56] propose an iterative procedure to arrive at appropriate taxes which requires none of the informational prerequisites of the Wellisz-Diamond optimal taxes. Following Davis and Whinston we will restrict attention to the case of two firms imposing reciprocal externalities on each other. Let the cost functions be

$$C^1 = C^1(x_1, x_2) \quad C^2 = C^2(x_1, x_2). \tag{5.19}$$

Introduce new variables, q_1 and q_2, which have the following interpretation: q_1 is the output of x_1 that firm 2 wishes firm 1 would produce. By the same token q_2 is the output of x_2 that firm 1 wishes firm 2 would produce. Now let the two firms solve the respective problems

firm 1's problem: firm 2's problem: (5.20)

MAX $\pi^1(x_1, q_2)$,   MAX $\pi^2(q_1, x_2)$.

x_1, q_2 x_2, q_1

If $x_1 \neq q_1$ and $x_2 \neq q_2$, then a tax can be obtained as a fraction of the difference,

$$t_i(0) = \delta(q_i(0) - x_i(0)), \quad 0 < \delta < 1 \tag{5.21}$$

$$i = 1,2.$$

With this initial (0th iteration) tax, let the firms solve the problems in (5.20) again and recompute the tax on the new solutions as follows:

$$t_i(1) = t_i(0) + \Delta t_i(1) \text{ where} \tag{5.22}$$

$$\Delta t_i(1) = \theta(q_i(1) - x_i(1)), \qquad 0 < \theta < 1$$

$$i = 1,2.$$

and feed this new tax to the firms and have them solve (5.19) again. When $q_i = x_i$, $i = 1,2$, the optimum is reached, and it remains for the authority to enforce the optimal transfers. The informational requirements are minimal, but there are transactions costs. Firms must spend resources on playing the iterative game, and the enforcing authority absorbs resources as well.

In order to know whether or not a prohibition or directive is optimal, the informational requirements are as great as they are for solving the centralized allocation problem. For while it is relatively easy to set a standard, it is not easy in terms of information to set the correct standard.

**Policy to Achieve
Optimality: Pollution
and Congestion**

Introduction

In this chapter voluntary and involuntary adjustments to a situation characterized by a pollution or congestion externality are explored. In all cases the
attention is focused on the level of pollution or congestion rather than the level
of output or consumption.

Voluntary Action: Bargaining and Merger

The voluntary actions explored here are merger, which applies only to firms, and
bargaining, which applies to any decision-making units. In general, the economics of bargaining and merger does not change from the case of nuisance to
the case of pollution. The difference centers only on the fact that the parties are
bargaining over the level of pollution and not over the level of the legitimate
activity (of production or consumption) which causes the pollution.

   Considering the bargaining solution first, we may note that if consumers are
involved, there are income effects in general. The fact that there are income
effects implies that the placement of liability has a short-run effect on the
outcome of the bargain. However, even in the case where there are no income
effects, one must qualify the Coase result that liability does not matter to the
extent that liability does matter if the functions involved are nonseparable.[1]
Moreover, since liability rules alter the distribution of rents, and since entry and
exit (long-run adjustment) occur in response to differential rent distributions,
the long-run equilibrium state of the world may be expected to look different
under one liability assignment than under another. This is true even though
equilibria under alternative liability assignments are optimal in a Pareto sense.

   Another problem is that bargaining is rarely a marginal process. As Wellisz
and Arrow point out, when a small number of units are involved in the
bargaining, there is no force propelling the bargainers to an optimum. There
exist a whole range of mutually profitable bargains which can be struck on finite
(nonmarginal) movements away from the status quo. Moreover, the exercising of
threats and other bargaining strategy decreases the chance that a bargain is
optimal. Finally, where large numbers are involved, for example urban air
pollution, the costs of organizing and determining the group will be likely to be
prohibitive so that no bargaining appears spontaneously.

In short, the considerations that apply to bargaining in the nuisance case also hold in the pollution and congestion case. Liability does matter in general despite the fact that any liability assignment might result in some optimal allocation. Moreover, an optimal bargaining outcome is most probable in the large numbers case, which is precisely the case for which prohibitive transactions costs are likely to be incurred. On the other hand, the small numbers case, where transactions costs are likely small enough to permit a bargain, is precisely the case where bargaining is least likely to achieve optimality. Only under a very restrictive set of circumstances can we expect bargaining to occur and the solution to be optimal.

The economics of merger is also the same in the pollution case as in the nuisance case. Merger can only occur between firms, so the merger solution leaves the case of pollution affecting consumers or consumers and firms unresolved. But in a purely voluntary setting, where no laws can force involuntary transfers, firms which are linked externally to the market by pollution can make a Pareto move through merger. Joint profits of the merged firms are larger than the sum of the profits of the individual firms. Therefore, some distribution of the stock of the new firm can be found that will be favorable to the stockholders of all the individual firms involved. However, in a legal environment that provides for involuntary transfers, such as through "nuisance" laws, trespassing laws, laws providing for injunctive relief, or torts laws, firms which have an incentive to seek a merger otherwise may seek a settlement through the courts which is more favorable to them than a merger would be. Whether or not a given firm goes to the courts depends on its subjective estimate of the expected value of a suit compared to the value of a merger, and on its own risk preference.

Involuntary Actions

The measures considered here are involuntary in the sense that economic units would not voluntarily choose to have these measures imposed given the alternative of no measures. This implies that the policy actions discussed here must be undertaken, if at all, by a collective decision-making unit such as a governmental "authority" established for the purpose, or some preexisting unit of government. The policy alternatives considered are charges and payments schemes, and prohibitions and directives.

Charges and Payments

Charges refers to the charging of a fee to those contributing to the pollution or congestion, while payments refers to paying those contributing to pollution or congestion to refrain from so doing. The first consideration is given to charges. From Chapter 4 we know that to achieve an optimum pollution reduction in a given shed with many pollution sources, it is necessary to equate the marginal cost of abatement for each source with the marginal external damages inflicted within the shed. Symbolically, the problem is to maximize the social product expressed as

$$\pi(s_1, \ldots, s_m) = B^1(s_1) + \ldots + B^m(s_m) \qquad (6.1)$$
$$- D(s_1 + s_2 + \ldots + s_m)$$

where B^i are the benefits from abatement costs avoided and D is the external damage inflicted. The B^i represent the abatement costs avoided if the intrafirm-efficient, or least-cost, abatement techniques are used. The necessary conditions for a maximum are that

$$dB^i/ds_i - dD/ds = 0, \qquad i = 1, \ldots, m, \qquad (6.2)$$

which implies

$$dB^1/ds_1 = dB^2/ds_2 = \ldots = dB^m/ds_m = dD/ds. \qquad (6.3)$$

This equilibrium is depicted in Figure 6-1 for the case where $m = 2$. A charges scheme imposes a per unit fee in the amount of the incremental damages at the optimum, $dD(s^o)/ds = v^o$, on the most immediate cause of the damages, the pollution itself.[2] Referring to Figure 6-1, if a unit charge of v^o is levied on pollution, then each pollution source demands on amount of pollution s_i^o and chooses to abate an amount $(\overline{s}_i - s_i^o)$. It chooses to abate precisely that amount because the marginal cost of abatement on those units of pollution is less than the unit charge. Abating beyond s_i^o costs the firm more than the unit charge, so it opts to demand s_i^o amount of pollution for which it pays $v^o \cdot s_i^o$ in pollution charges. If the quasi-demand functions for pollution represent the minimum abatement costs avoided under intrafirm-efficient abatement techniques, then the charges scheme provides not only for the socially optimal aggregate amount of pollution to be demanded, but it also provides the proper incentives so that the distribution of abatement among firms that is achieved is such that no other distribution can achieve the same aggregate reduction at a lower cost. In short interfirm efficiency is achieved. Moreover, the charges scheme creates the proper incentive for achieving intrafirm efficiency in abatement as well. A firm facing a charge of v^o per unit of effluent always stands to save in charges by abating (and demanding pollution) according to a schedule representing the most efficient abatement techniques (see Figure 6-2). Another advantage of the charges scheme is that once the scarcity value of pollution in a given shed is set, the abatement decision-making is decentralized to the various pollution sources.

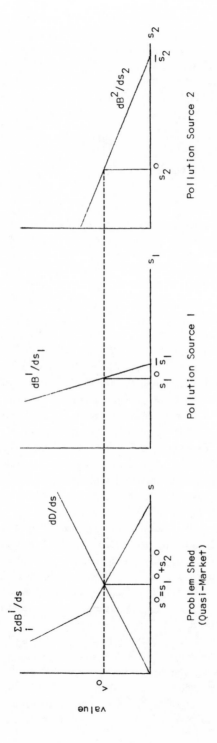

Figure 6-1.

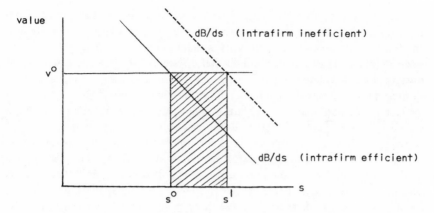

Figure 6-2. Saving in Charges with Intrafirm-Efficient Abatement.

The charges scheme for obtaining optimality runs into trouble most notably from Coase. He argues that

It is enough for my purpose to show that, even if the tax is exactly adjusted to equal the damage that could be done to neighboring properties as a result of the emission of each additional puff of smoke, the tax would not necessarily bring about optimal conditions. An increase in the number of people living, or of businesses operating, in the vicinity of the smoke-emitting factory will increase the amount of harm produced by a given emission of smoke. The tax that would be imposed would therefore increase with an increase in the number of those in the vicinity. This will tend to lead to a decrease in the *value of production* [my emphasis] of the factors employed by the factory, either because a reduction in production due to the tax will result in factors being used elsewhere in ways which are less valuable or because factors will be diverted to produce means for reducing the amount of smoke emitted. But people deciding to establish themselves in the vicinity of the factory will not take into account this fall in the value of production which results from their presence. This failure to take into account costs imposed on others is comparable to the action of a factory-owner in not taking account of the harm resulting from his emission of smoke.[3]

The conclusion usually drawn from Coase's argument is that not only must the firm be taxed to take into account the external damage caused by its smoke, but also the residents must be taxed to take into account the costs internally imposed on the firm.[4]

There are three points that need to be made with regard to the Coase quote. First, as noted in Chapter 4, the supply of pollution (damages) shifts with the entry of new residents or potentially damaged businesses into the region. This implies, as Coase points out, that the optimal scarcity value of a unit of smoke increases as new residents move in. But residents deciding where to move will choose to move to that location, other things equal, where the damage from pollution is the smallest. Therefore the damages of a new resident moving into an air basin represent a genuine scarcity which should be reflected in the scarcity value of smoke. There is no fundamental difference between the new resident shifting the supply curve of pollution, and hence increasing its price, and his shifting the local supply curve of labor, and hence lowering its price. The increase in the unit charge for pollution due to the new resident's presence is a pecuniary externality and not a technological one, and as such it does not represent a misallocation anymore than does the falling of the price of labor.[5]

The second point is that the value of production of the smoke-emitting firm may or may not fall. It may, in fact, rise. If a firm faces a unit charge on its smoke, and the charge increases due to new residents moving into the basin (because a move elsewhere would cause them to incur greater damages), then the marginal cost of output increases. This in turn causes the output price to increase which causes the firm to adjust its output level with regard to the new price and marginal cost function. But since the value of output is price times quantity, the new value of output after the increase in the unit charge on smoke may be either greater or smaller than before the charge increase. But whether or not the value of production rises or falls is irrelevant. The important point in a neoclassically flexible situation is that the increase in the unit effluent charge causes the firm to adjust its use of smoke as an input, that the increase in the cost of production be transmitted to the market through altered supply conditions (marginal cost), and that ultimately consumers face a higher price (given constant demand) for the output.

Third, the notion that both the residents in a basin and the smoky factory need to be taxed for optimality is wrong. Baumol demonstrates in the context of a nuisance model that a tax on output (of the nuisance-causing good) is all that is required for optimality. This demonstration can be seen by referring to the optimality conditions in Chapter 3 (models 1, 2, and 3) and by noting that taxing the nuisance-causing good in the amount of the external marginal harm (either utility or cost) causes optimality to be achieved.[6] Baumol emphasizes that "... the solution calls for neither taxes upon [the externality victim] nor compensation to [it] for the damage it suffers."[7] Moreover, the Pigovian tax is precisely the vehicle needed to achieve optimality. It provides for "... a nonzero price (tax) to the supplier of the [externality], and a zero price to the consumer,"[8] which is exactly what the optimal price should be for a public externality. The appropriate price to the consumer is zero because of the nonexclusiveness of the externality (smoke), while the appropriate price to the supplier is positive because his activity imposes a cost on society's resource base.[a]

---

[a]Coase, in private correspondence to Baumol (see W.J. Baumol, "On Taxation and the Control of Externalities," *American Economic Review* 62 (June 1972): 314-15), indicates that the point he is trying to make is that once the optimal Pigovian tax is computed (in

While Baumol's demonstration is in terms of a nuisance model, the extension to the pollution case is obvious. The optimality conditions in a problem shed call for the equation of marginal abatement costs for each pollution source to the common marginal damage level. An appropriately set price on the effluent causes the margins to be equated for optimality. The tax is not a price in the ordinary sense because it is really only a half-price: That is, the fee is paid by the demander of pollution, but it is not (and need not for optimality[b]) received by the supplier of pollution (the damaged victims). In addition, it remains true that as long as those seeking a residential location choose a basin in which the damages are smallest, other things equal, the damages of a new resident represent a genuine cost to society.[9] This cost should be reflected in the charge for pollution within the shed, and ultimately should cause output price to increase through altered cost and supply conditions.

The payments scheme consists of making per unit payments to the source of pollution for agreeing to refrain from discharging units of pollution. With reference to Figure 6-1, the argument is that if a payment of v^O per unit of effluent reduction is announced, the sources of pollution will respectively reduce their pollution level to s_1^O and s_2^O. They are induced to do so because the payment on each pollution unit between $\bar{s}_i$ and s_i^O is greater than the marginal cost of abatement, which means a return net of abatement costs is realized on those units of pollution. But beyond s_i^O abatement is more expensive than the unit payment, so the pollution source is induced to demand s_i^O pollution.

While it has been argued that the system of payments is fully equivalent to the system of charges,[c] there are problems with this propositon. The first problem is that there is no natural origin from which to measure reductions in pollution.[10] The correct origin, of course, is $\bar{s}_i$, the level of pollution that would have been produced in the absence of any intervention at all. But if there are payments for reduction in pollution, the incentive for the polluter is to discharge as much pollution as possible so that the "origin" from which his reductions are measured is as large as possible. Once the payments system is announced, there is no way, short of having an omniscient taxing authority, to know what a given polluter would have done in the absence of any policy. The origin for payments must by default be the current amount being discharged, for which every

static conditions), it need not be changed if there are nonoptimal departures from equilibrium, as when residents move. However, this is not the common interpretation given to Coase. If, on the other hand, enough information is not available to calculate the optimal Pigovian tax, Baumol shows that, under standard assumptions about an adjustment mechanism, setting the tax equal to the current nonoptimal damages results in an optimal solution after adjustment. See Baumol, ibid., p. 315.

[b]Again, when bargaining is not economically feasible.

[c]R. Coase, op. cit., has reference mainly to legal liability, but the extension to tax liability seems immediate. An argument based on Coase holds that the allocation of resources is insensitive to the assignment of tax liability (i.e., whether there are payments or charges). See also A.V. Kneese and B.T. Bower, *Managing Water Quality: Economics, Technology, and Institutions* (Baltimore: Johns Hopkins University Press, 1968), p. 101ff.

incentive exists to make it larger than it would have been without any policy. Thus Kamien, Schwartz, and Dolbear show that the amount of pollution discharged is greater under a payments scheme, using the current amount being produced as the origin for the payments, than it is under a charges scheme. Indeed, the amount might be greater than the amount that would have occurred without any intervention at all because of the perverse incentives.[11] As a result of this Freeman concludes that

An authority should not place itself in the position of having to base its action (in this case a bribe schedule) on information that may not be available (in this case the maximum waste discharge levels, W^* [that would occur without intervention]) if it can achieve the desired results using a more limited set of available data (in this case the incremental damage function).[12]

The second problem with the proposition that the allocation of resources is invariant with respect to tax liability is that whether a polluter is paid or charged affects its profit position, which in turn affects the long-run attraction of resources to the polluting activity.[13] While it is true that some optimal outcome may be expected under a payments scheme in the sense that all Pareto relevant externalities have been eliminated, it is not in general true that that optimal outcome is the same one that occurs under a charges scheme. And the reason is that the long-run industry and output mix varies according to which scheme is used because of the differential attraction of resources to the polluting industry.

There are additional considerations when weighing a charges scheme against a payments scheme. The first is convention and our sense of equity.[14] If we think of the traditional problem of allocating labor and goods, is clear that there, too, there is a choice between charges and payments. The conventional charges way to view this problem is that producers are charged by labor for their services while labor is charged by producers for the goods they purchase. It is equally possible to run an economy of this simple type on the basis of payments. Under this scheme, producers have claim to the services of labor at all times except for the leisure time specifically purchased from the producers at a per hour rate. On the other hand, labor may "have" the quantity of goods it wants except for a quantity it has specifically agreed not to demand in return for a per unit payment. Under this system, a supply and demand for leisure and a supply and demand for abstaining from taking goods would evolve. These market forces would allocate labor and output optimally in the Pareto sense, but this scheme seems outrageous to us in the context of allocating private goods and services because of the way that economic and legal conventions have evolved. Moreover, the problem of ascertaining from what origin payments should be made for restricting demand dramatically presents itself. In the absence of an omniscient planning authority, each consumer's word must be taken. But it is certain that not a few among our number would deign to accept per unit payment for restricting our "demand" of six running Jaguars, five sailing yachts, four gliding planes, ..., and a trip to the tropical sun. Therefore, the convention that demanders of scarce resources should be charged rather than demanders of scarce resources being paid for restricting demand, and our sense of equity underlying that convention, favor the charges over the payments scheme.

The next consideration when thinking about payments versus charges is that payments must continue to a firm that has found it profitable to shut down, thereby reducing its effluents to zero. Payments must also be made to firms that would potentially enter the basin were it not for the payments.[15] These problems present snags in the administration of the payments scheme that are not presented in a charges scheme.

Prohibitions and Directives

As with the nuisance case, prohibitions and directives can be used to control externalities. In this case the prohibition or directive concerns the level of pollution or congestion, and not the level of the legitimate activity causing the pollution. In an environment of certainty, an outright prohibition on a given type of pollution within a problem shed is optimal when the marginal damage cost of the pollution exceeds the marginal abatement costs at each level of pollution. An outright prohibition is also optimal in the static sense if the presence of even the smallest amount of pollution causes irreversibilities for which an option demand[d] exists which exceeds the marginal abatement costs of the last unit of pollution. For example, if the use of even the smallest amount of a given pesticide causes the loss of an animal species, which is an irreversible event, and if the option demand for the preservation of the species exceeds the demand for the first unit of the pesticide as an input,[e] then a prohibition of the pesticide use is optimal.

Directives may take the form of a requirement or a standard, where the difference is the degree of choice permitted. A requirement permits less choice than a standard but more choice than a prohibition, which affords no choice at all.

[d]Option demand is the willingness-to-pay for the option to exercise effective demand for something at some future time, whether or not the option is actually exercised.

[e]The function dB/ds gives the marginal costs avoided from not abating at each argument s. These costs depend on how other inputs substitute for s in the production process, and they therefore give the demand for s, as noted earlier, so that the function dB/ds is really an input demand function. See M.L. Langham, J.C. Headley, and W.F. Edwards, "Agricultural Pesticides: Productivity and Externalities," in A.V. Kneese and B.T. Bower (eds.), *Environmental Quality Analysis* (Baltimore: Johns Hopkins Press, 1972), pp. 181-212, who estimate the elasticity of substitution of pesticides for cropland. They find a relatively large elasticity (a 12 percent increase in cropland would compensate for a 70-80 percent reduction in chemical pesticides), which means that if the actual price ratio of pesticides to land understates the true ratio because of externalities by even a small amount, a relatively large overuse of pesticides results.

Requirements most often take the form of prescribing that a certain device or process must be used, which is in effect dictating a portion of production technology. For example, all automobiles are required to have certain devices for the control of emissions. The problem with a requirement of this type is that there is no mechanism insuring interfirm, or in this case intercar, optimality which requires that the marginal cost of abatement be equated for each pollution source. Requiring device A, B and C on all cars equates marginal abatement costs only by accident if at all.

A second problem with a requirement prescribing a device or process is that there is no assurance that the required action is the intrafirm least cost way of doing things. For example, Kneese lists twenty-five ways of treating industrial and municipal water borne wastes.[16] A requirement that a pollution source along a river use any specified one or combination of these processes would only by the merest luck result in the cost minimizing technology for pollution control within the firm.[17]

Further problems with a directive requiring specified behavior center on incentives. First, there is no incentive to do any better than the requirement indicates even if it is possible.[18] If changing technology permits the improvement of pollution control, there is no mechanism insuring the new technology will be used short of another directive requiring its application. Moreover, even with constant technology, increased application would reduce pollution further. But since doing so involves only costs and no benefits to the firm, there is no incentive to undertake action in excess of the required action. Second, there is no incentive for the polluting firm to develop new more efficient control technology because such a development would not give the firm a competitive advantage over its polluting competitors. Nor is there any market incentive for research and development in pollution control outside of the firm. Since firms need only meet the requirement and nothing more, no entrepreneur could hope to market a new process or device short of having the authority require his device over some other.

The setting of standards represents the second type of directive which may be used to control pollution type externalities and may be either point-by-point discharge standard or an across-the-board standard.[19] For example, an across-the-board standard may take the form of directing that a given absolute amount or a given percentage amount of the pollutant be eliminated by all sources in the basin. One problem with this approach is that whatever level of ambient quality improvement is realized, an even greater improvement could be realized for the same cost outlay. This is because in general the across-the-board standard in either absolute or percentage terms fails to meet the marginal conditions for efficiency in pollution reduction (see (6.3) and Figure 6-1). A point-by-point standard treats each individual pollution source individually. If rational distinctions are to be made among pollution sources, the differences in marginal abatement costs must be taken into account. But this means that the basinwide authority must know the demand for pollution function for each source in order to set a rational standard for each source. This is a giant administrative undertaking which involves large costs so that in order to justify a point-by-point standard scheme, the allocative gains must exceed these administrative costs. A problem common to either standard setting scheme is the problem of incentives encountered above. There is never any incentive for the polluter to do any better than the standard directs because such an action imposes costs on, but secures no benefits for, the polluter.

There is a situation in which setting standards works while levying charges does not. That concerns the case where the optimal charge is less than the average cost of abatement. If that occurs, a firm facing the optimal charge of v^O would choose to pollute in the amount $\bar{s}$ paying a fee $v^O\bar{s}$ rather than abate an amount $(\bar{s} - s^O)$. This is because abating to the optimal level, s^O, involves costs $((AC^O(\bar{s} - s^O)))$ which are greater than the charge that would be levied in the absence of abatement. The firm stands to lose an amount depicted as the shaded region in Figure 6-3 if it abates. So a maximizing firm will simply pollute at its maximum level, $\bar{s}$, and bear the charge levied against it. In this case, Kneese and Bower point out that a standard directing the polluting firm to restrict its pollution to s^O achieves the optimal discharge whereas a charge scheme cannot.[20] The difficulty of precisely determining the optimal level of effluent

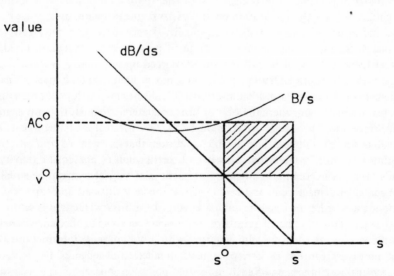

Figure 6-3. Source: This figure is a modification of the one appearing in A.V. Kneese and B.T. Bower, *Managing Water Quality: Economics, Technology and Institutions*, (Baltimore: Johns Hopkins Press for Resources for the Future, 1968), p. 138.

discharge for each pollution source increases with the number of sources in the basin, but even an across-the-board standard may be a Pareto improvement over either the case of permitting unbridled pollution or of levying a charge not sufficiently high to induce some abatement.

Integration of Approaches

The difficulty and expense of literally applying a charges scheme to the control of pollution leads some authors to propose a mixing of approaches in dealing with the problem. Some of these difficulties involve the application of physical models in determining exactly what the damages are. For example noise propogation models, air diffusion models, and models such as the Streeter-Phelps equation of oxygen sag in water are needed if damages are to be accurately assessed.[21] Moreover, even with accurate physical models, data on the nature and extent of damages due to various pollution types is often not available and is in fact difficult to obtain. These problems involved in determining a precisely optimal charge for effluents lead some writers to advocate setting a more or less arbitrary standard for ambient quality within a problem shed and then using market-like mechanisms to achieve the standard. The standard may be the result of the political interaction of various interested pressure groups, or it may be the result of an arbitrary decision on the part of a regional authority using the best information available. The decision as to the appropriate standard would also hopefully involve the application of the best available estimates on damage costs and the costs of achieving different quality levels. After the setting of the standard, Baumol[22] and Baumol and Oates[23] propose that a charge be levied on the effluent and that it be raised until the agreed upon standard is met. In this way, even if the optimal level of pollution is not secured, whatever level is finally reached is achieved at a minimum of cost. This is because each pollution source abates to the margin where the cost of further abatement equals the unit charge, and, since the unit charge is common to all, the marginal abatement costs are equated for all sources. This is the condition that is required for interfirm optimality which insures that a least cost distribution of abatement activity is undertaken. The common marginal abatement cost may or may not be equal to the marginal damage cost, and in fact it may be very difficult and expensive to determine whether or not it is, which is why the arbitrary standard is set in the first place. This mixture of standards and pricing secures the efficiency benefits of the market and can be administrated with a minimum of information. It is not necessary for the authority to have detailed knowledge of the marginal abatement cost functions of each individual pollution source, nor is it necessary to have the last word on marginal damages.

A similar approach is put forward by Dales,[24] who proposes that the notion of property rights be altered from what they are now. Instead of everyone having the right to use a common environmental medium, such as air and water, as he sees fit, no one would have the right to pollute unless those rights are specifically purchased. The regional authority sets a standard and issues rights in that amount with an attached expiration date. Then no one could legally pollute without ownership of a right (defined in terms of some appropriate physical quantity). A market is then established in pollution rights, insuring that the price reflects the basinwide marginal abatement cost, that the rights are efficiently distributed, and that abatement takes place efficiently. Again this combination of standard setting and pricing secures all the efficiency benefits of a market and economizes on the information needed for administration.

Roberts[25] and Kneese[26] argue for a basinwide authority to deal with problems of water pollution. This authority would have complete jurisdiction over all matters concerning the water basin, and should be the only one permitted to dump waste into the water course. Firms, municipalities, or individuals wishing to dump waste would be charged according to the marginal damage within the basin after optimal treatment. The authority should ideally run all treatment plants to capture any economies of scale in treatment and to permit optimal location of the treatment facilities. The authority would also use prohibitions, and standards coupled with its charges mechanism. For example, the authority might determine that stream specialization is warranted within the basin, the application of which is an example of separate facilities used in the control of a nuisance.[27] One stream might be preserved in a relatively pure state for fishing and swimming while another might be given over entirely to waste disposal. The *Genossenschaften* of the Ruhr Valley in West Germany are associations of the type which closely correspond to the notion of a basinwide authority.[28] They enforce stream specialization, operate treatment plants, and impose effluent charges on polluters in the basin. This integrated approach using charges, standards, prohibitions when needed, and separate facilities is a realistic approach in dealing with real world problems. But both Roberts and Kneese and Bower are convinced that effluent charges have an important part to play in the control of pollution.[29] Some method must be used to get polluters to realize as an opportunity cost to them the external damage caused by their actions, and the charges method is a way to do that which is not administratively impossible. The idea of the basinwide authority extends naturally to the case of air pollution as well.

Information Requirements

Application of a literal charges scheme, as well as efficient standards or requirements, necessitates full knowledge of the damages function and the individual marginal abatement cost functions. Without knowledge of the individual marginal abatement cost functions, the authority could not aggregate to

122

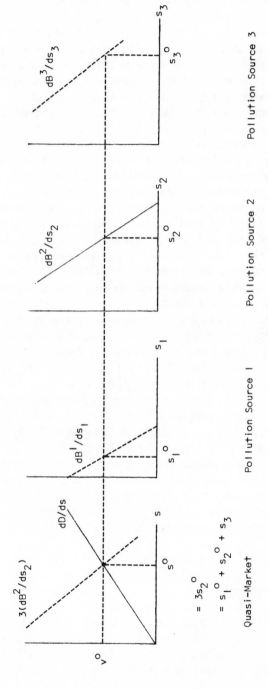

Figure 6-4. Source: Adapted from A.V. Kneese and B.T. Bower, *Managing Water Quality: Economics, Technology, and Institutions*, (Baltimore: Johns Hopkins Press for Resources for the Future, 1968), p. 137.

derive the basinwide demand for pollution function. This is a very expensive and difficult process and is tantamount to solving the centralized planning problem. If the marginal abatement cost functions are linear or can be closely approximated by linear functions, Kneese and Bower point out that substantial economies in informational requirements result.[30] Rather than needing knowledge of each individual pollution source's marginal abatement cost function, only the average function is needed for the application of an optimal charges scheme. For example, in Figure 6-4 if the pollution demand function for source 2, dB^2/ds_2, is the average of sources 1 and 3, then in order to establish an optimal charge of v^O, the authority must know the marginal damages function, dD/ds, and the average demand for pollution function, dB^2/ds_2. This information economy is not available if the authority wishes to impose efficient standards on the pollution sources, however. Efficient standards can only be determined with full knowledge of the pollution demand function for each source as well as the damages function whether or not these functions are linear. But even with a large number of pollution sources in the basin so that sampling must be undertaken to estimate the average demand for pollution, there accrue substantial informational, and hence, cost, advantages to a charges scheme over setting optimal standards point-by-point.

If we retreat from maximizing behavior under the duress of transactions costs, and settle for satisficing behavior in a given problem shed following Baumol and Dales, substantial information advantage is gleaned. Using the best information available on the costs of pollution, an authority can set a basinwide standard and either issue pollution rights in that amount[f] or impose a charge sufficiently high to achieve the standard. In either case, no knowledge of individual demands for pollution need be known. Either of these two systems of control is superior to both across-the-board standards (for efficiency reasons) and point-by-point standards (for informational cost reasons) in achieving given basinwide standards.

[f]This, of course, requires an accommodating change in the legal system.

7

Theory of External Effects: General Equilibrium and Dynamic Optimality

General Equilibrium

In Chapter 3 through Chapter 6 the application of partial equilibrium analysis to the externality problem is made. Despite the fact that in Chapter 3 optimizing techniques are applied to an entire, hypothesized, simple economy, the examination of the externality problem is carried on under the assumption that the rest of the economy is running along optimally. For example in the case of nuisance in consumption, the production side is ignored, and the good which does not create a nuisance is assumed to be optimally allocated among consumers. In the pollution case, it is even more clear that the analysis is partial equilibrium in nature. Attention is focused on the optimal use of a given problem (air, water, or noise) shed with respect to a given flow of pollution or congestion, which we labeled s. The shortcoming of the partial equilibrium approach is the same in nonmarket as in market allocative problems, and that is that substitution, complementarity, and linkage effects of all types among markets are ignored. For example, incineration of solid waste substitutes airborne waste disposal for landfill waste disposal, or water scrubbing of smoke stack discharges substitutes waterborne for airborne disposal. Solutions that appear to be rational in a partial equilibrium framework may turn out to be irrational on a more general consideration because the "all other things equal" assumption on which the partial equilibrium approach is based is in general not true as adjustments in any one market are made.

A number of writers have approached the nonmarket problem from the general equilibrium point of view including Leontief,[1] Ayres, Kneese, and D'Arge.[2] Leontief extends the input-output approach that he developed in order to handle the undesired residuals that do not pass through a market. Ayres, Kneese, and D'Arge extend the Walras-Cassel general equilibrium model to include nonmarket scarcities. These writers are instrumental in changing the attitudes of economists from believing that externalities problems represent a minor, occasional abberation in the allocative mechanism to realizing that residuals are an integral and pervasive part of the production and consumption process. Russell and Spofford present what they term a management model of a basin, or region, which is general to that region in the sense that it accounts for all of the linkages affecting residuals handling, all of the interindustry linkages, and all of the damages to all receptors from all sources through all media in the basin.[3] Conceptually their model is perfectly general to a region even if in practice expenditure limitations necessitate abstraction and simplification. Of course, its regional character precludes the consideration of interregional linkages, which makes it less than a completely general model, but it is sufficiently general and sufficiently well developed for actual application that it warrants special consideration.

The Russell-Spofford model consists of three submodels: a model of production and residuals handling, an environmental model of residuals diffusion, and a model of receptor damages.[a] The model of production and residuals handling is a linear programming model with eight sets of activities, four sets of constraints, and an objective function. The eight activities sets are production, sales, imports, by-product extraction, raw material recovery, treatment of residuals, transport of residuals, and discharge of residuals. The constraint sets are conditions placed on production and sale, input availability, primary residuals, and secondary residuals (that is, residuals generated while treating residuals). They are specifically of the following form: (1) all sales must be either produced or imported, (2) input use must be less than or equal to input availability, (3) all residuals generated must be accounted for by residuals handling (B, W, T, V) and by residuals discharge (R), and (4) a similar constraint for secondary residuals. Writing the vector of activities as $[X, Y, M, B, W, T, V, R] = Z$ and the vector of costs or prices per unit of activity as $[c_x, -p_b, c_w, c_t, c_v \; t] = c$, we can represent the production and residuals model as standard programming problem:

minimize: $c^1 Z$ (7.1)

subject to: $AZ \geqslant b$

$Z \geqslant 0.$

The second of the Russell-Spofford submodels is the environmental diffusion model. It uses as input the residuals discharged in the production model and transforms them into ambient concentrations at various receptor locations. The transformation takes place through a matrix of transfer coefficients that derive from physical diffusion models. The receptor locations are map grids in the case of air pollution, or river segments in the case of water pollution. The ambient concentrations are given by

$C = QR$ (7.2)

where R is the level of residuals discharged in the production model, and Q is a matrix whose typical element, q_{ij}, gives the increase in the residual concentration at receptor location i as the result of a unit discharge on location j.

[a]The discussion of the Russell-Spofford model which follows is developed from the author's understanding of the cited sources. No attempt is made to cite those works in what follows, with the understanding that the ideas discussed are Russell and Spofford's.

Table 7-1
Linear Programming Model of Production and Residuals Handling

	Production Alternatives X	Sales & Imports Y, M	Raw Material Extraction B	Raw Material Recovery W	Treatment & Transport of Residuals T, V	Discharge of Residuals R	RHS
Production and Sale							
Input Availability							$\geq b$
Primary Residuals							
Secondary Residuals							
Objective	c_x Cost of Production	p Prices	c_b Cost of Extraction	c_w Cost of Recovery	c_v, c_t Cost of Treatment and Transportation	t Effluent Charges	

Source: C.S. Russell and W.O. Spofford, "A Qualitative Framework for Residuals Management Decisions," in A.V. Kneese and B.T. Bower, eds., *Environmental Quality Analysis* (Baltimore: Johns Hopkins Press for Resources for the Future, 1971), p. 141.

The final submodel relates the ambient residual concentrations to damages,

$$\mathbf{D} = f(\mathbf{C}) \qquad\qquad\qquad (7.3)$$

That is, the ambient concentration of each residual is related to damages in each receptor location. From this information marginal damages can be calculated.

The complete model works together in an iterative fashion. To start, the production model is optimized with $t = 0$ in the objective function; that is, effluent charges are nonexistent. This will generate a first stage optimal activity level for R, the residuals discharge, taking into account all the tradeoffs among treatment modes and secondary residuals (i.e., stack-scrubbing causes water pollution, etc.). This residuals discharge level is pumped into the environmental diffusion model which reads out ambient receptor concentration levels. These levels in turn are used in the receptor damage model to determine damages, and specifically marginal damages,[b] of each residual at the various receptor locations. These marginal damages are used as effluent charges in the next iteration of the production model. This process is continued until there are no more improvements in the objective function. The solution gives us, among other things, the optimal level of residuals discharge of each type, taking into account all of the linkages among residuals and their reduction and the associated costs. In other words, it gives the general equilibrium level of residuals discharge in the region.

The advantages of the Russell-Spofford approach over the partial equilibrium approach of considering a given basin with respect to one given pollutant are many. The more general approach takes into account (1) the relative residual intensiveness of production of alternative industries, (2) the tradeoffs that exist in handling residuals, such as recovery, treatment, discharge, (3) the interrelationship of residual types, as when "treating" air pollution causes water pollution, that is the problem of secondary residuals, and (4) the marginal damage of receptors by imposing effluent charges on the discharge of residuals. The disadvantages are not problems with the conceptualization of the model; that is impeccable. All of the disadvantages lie in practical considerations. As Dorfman points out, work on the third submodel level, that of determining receptor damages as a function of residual concentrations, is in a painfully underdeveloped state.[4] The data that are required for that type of modeling is often not available, and those that are available often underestimate damage because of the intangible character of the damages caused by some residuals. As indicated in Chapter 6, the expense of applying a model with the degree of sophistication of the Russell-Spofford model may outweigh any misallocation that results from simply setting arbitrary ambient quality standards and applying effluent charges sufficiently high to achieve the standard.[5] But this is not to shadow the advantages of the more precise general approach when transactions costs are low enough.

---

[b]If specific damage functions are known, marginal damages can be determined analytically; otherwise numerical determination can be made.

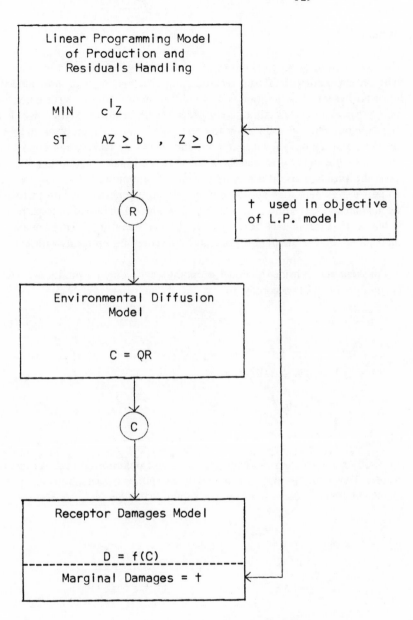

Figure 7-1. Schematic of Russell-Spofford Iterative Model. Source: C.S. Russell and W.O. Spofford, "A Qualitative Framework for Residuals Management Decisions," in A.V. Kneese and B.T. Bower, eds., *Environmental Quality Analysis*, (Baltimore: Johns Hopkins Press for Resources for the Future, 1971), p. 123 modified.

Dynamic Models

The examination of externalities under static conditions is useful in bringing to light certain modes of thought regarding the problem. As with most modeling, however, the static assumption is a simplification of reality undertaken to focus more clearly on certain allocative aspects of the problem. Static analysis has shortcomings when a reallocation to correct an externality problem involves an intertemporal change of consumption, and when the reallocation takes time.[6] The longer the time taken, the less acceptable is the static assumption. We could solve the problem by attaching the words "discounted value" to costs and benefits everywhere they appear, but models more explicitly dynamic which use the mathematics of control to examine continuously optimal time paths for a system with externalities have surfaced in the literature.[7] These models are applications of the Maximum Principle[8] to economic problems with externalities.

For example, consider a model economy where directly productive capital, k_1, produces output[c] according to

$$x = f(k_1) \tag{7.4}$$

where

$$f' > 0, \; f'' < 0, \; f(0) = 0 \tag{7.5}$$

$$\lim_{k_1 \to 0} f' = \infty, \; \lim_{k_1 \to \infty} f' = 0.$$

But producing output causes damages because of pollution from the production process. The pollution may be abated so as to reduce damages, however, and the abatement takes place with abatement capital, k_2. Therefore, damages are given by

$$z = g(k_1, k_2) \tag{7.6}$$

where

$$g_1 > 0, \quad g_2 < 0 \tag{7.7}$$

$$g_{11} < 0, \; g_{22} < 0, \; g_{12} = g_{21} < 0$$

$$\lim_{k_2 \to \infty} g_2 = 0$$

---

[c]Assume the labor force is constant so that we can abstract entirely from consideration of the labor variable.

Net output is given by gross output minus damages,

$$y(k_1, k_2) = f(k_1) - g(k_1, k_2). \tag{7.8}$$

In this model the damages occur to current output, but they might occur to the stock of capital, the stock of something nonproduced called the stock of nature, or to all three.[d] In the present formulation, initial conditions, $k_1(0)$ and $k_2(0)$,[e] are chosen so that

$$y(k_1(0), k_2(0)) > 0 \tag{7.9}$$

in order to have the problem make economic sense,

The object of production is consumption

$$c = y(k_1, k_2) - I_1 - I_2 \tag{7.10}$$

where I_i is investment in type i capital, or

$$\dot{k}_1 = I_1 - \delta_1 k_1, \quad \dot{k}_2 = I_2 - \delta_2 k_2, \tag{7.11}$$

$$\delta_i > 0, \quad i = 1,2.$$

The rates of depreciation are given by δ_i, and consumption is evaluated by a community utility function

$$U = U(c) \tag{7.12}$$

with the properties

$$U' > 0, \ U'' < 0 \tag{7.13}$$

$$\lim_{c \to \infty} U' = 0, \ \lim_{c \to 0} U' = \infty$$

Letting s_2 be the saving ratio of the economy and s_1 be the proportion of saving going to investment in capital of type one, we get

$$I_1 + I_2 = s_2 y(k_1, k_2) \tag{7.14}$$

$$I_1 = s_1 s_2 y(k_1, k_2)$$

$$0 \leqslant s_i \leqslant 1.$$

[d]See the current literature in this area for various ways of modeling the nondesirable property of pollution.

[e]Since all variables are functions of time, the time argument is ordinarily omitted.

The problem is to choose s_1 and s_2 so as to maximize an objective functional. Stated formally, the problem is to

$$\text{maximize: } V = \int_{O}^{\infty} U(c) \exp(-\rho t)\, dt, \quad \rho > 0, \qquad (7.15)$$

subject to (7.11) and (7.14) and initial conditions $k_1(0) = k_{10} > 0$ and $k_2(0) = k_{20} \geqslant 0$, where c is given by (7.10).

This problem of heterogeneous capital goods has been investigated in some of its aspects by various writers. Hahn analyzes the path that an economy takes with heterogeneous capital goods when labor does not save and capitalists do not consume.[9] His conclusions are that the problem of uniqueness is more pronounced than in the two-sector case, and that even if there is a unique path describing the development of such an economy, there are a wide range of initial conditions and parameter values for which the actual path does not converge to the equilibrium path. The Hahn model is consistent with a decentralized system where labor and capital earn factor payments related to their productivity. Samuelson investigates the "terminally optimal" problem of efficiently accumulating heterogeneous capital goods with an arbitrarily fixed saving ratio.[10] His analysis agrees with Hahn's conclusions on uniqueness, and he goes on to speculate about the role of perfect futures markets in coping with the Hahn problem of convergence to an equilibrium path when such a path exists. The paper by Samuelson and Solow[11] extends the Ramsey[12] problem to the case of many capital goods. By introducing "bliss" (saturation) points in either consumption or production (or both) they are assured of solely interior extrema which maximize a Ramsey-type objective integral. The optimum can be found by application of the classical calculus of variations. Problem (7.15) under investigation here is more easily handled by use of the Maximum Principle because of the inequality constraints and because of the possibility of boundary extrema when "bliss" points are not postulated.[13]

Application of the Maximum Principle

To apply the Maximum Principle, form the Hamiltonian function

$$H(\mathbf{k}, \lambda, \mathbf{s}) = \qquad (7.16)$$

$$U[y(k_1, k_2) - s_2 y]\exp(-\rho t) + \lambda_1[s_1 s_2 y(k_1, k_2) - \delta_1 k_1]$$

$$+ \lambda_2[(1 - s_1)s_1 y(k_1, k_2) - \delta_2 k_2]$$

where λ_1 and λ_2 are the dynamic shadow prices of investment in the two respective types of capital. Using the substitution

$$\lambda_i = q_i \exp(-\rho t), \quad i = 1,2, \qquad (7.17)$$

the Hamiltonian can be written as

$$H(\mathbf{k}, \mathbf{q}, \mathbf{s}) = \tag{7.18}$$

$$\left\{ U\left[(1 - s_2)y\right] + q_1\left[s_1 s_2 y - \delta_1 k_1\right] \right.$$

$$\left. + q_2\left[(1 - s_1)s_2 y - \delta_2 k_2\right] \right\} \quad \exp(-\rho t).$$

The necessary conditions for the state variables, $k_1(t)$ and $k_2(t)$, and the control variables, $s_1(t)$ and $s_2(t)$, to represent a solution to the control problem (7.15) are as follows.

First, the dynamic shadow prices must satisfy the adjoint equations

$$\dot{\lambda}_i = -\partial H / \partial k_i. \tag{7.19}$$

Second, s_1 and s_2 must maximize H, or

$$H = \underset{s_i \epsilon S}{\text{MAX}} \; H(\mathbf{k}, \mathbf{q}, \mathbf{s}). \tag{7.20}$$

where the control region S is the unit square in $s_1 \times s_2$ space. In addition, an optimal solution must satisfy the boundary conditions and the laws of motion of the state variables, (7.11). Deriving the adjoint equations in terms of q we get

$$\dot{q}_1 = (\rho + \delta_1)q_1 + \left[(q_2 - q_1)s_1 - q_2\right]s_2 \tag{7.21}$$

$$- (1 - s_2)U' \; (f' - g_1)$$

$$\dot{q}_2 = (\rho + \delta_2)q_2 + \left[(q_2 - q_1)s_2 - q_2\right]s_2$$

$$- (1 - s_2)U' \; (-g_2).$$

Interior Solutions. In the case of solutions interior to the control region we can look for a maximum of H by allowing the first partials of H with respect to s_i vanish, or

$$\partial H / \partial s_1 = (q_1 - q_2)s_2 (f - g) = 0 \tag{7.22}$$

$$\partial H / \partial s_2 = \left[(q_1 - q_2)s_1 + q_2 - U'\right](f - g) = 0.$$

There are two cases; the uninteresting case where $f = g$ and net output $y = 0$, in which case the choice of controls does not matter; and the case where $f - g = y > 0$. Since initial conditions are chosen to insure $y(0) > 0$, and since at anytime $t > 0$ for which $f = g$, capital is decumulating at the rate $\dot{k}_i = -\delta_i k_i$, we are assured that the singular case cannot be sustained for any finite time period.

In the nonsingular case where $y > 0$, we can derive from (7.22) that

$$q_1 = q_2 = U' \tag{7.23}$$

is necessary for a solution. This interprets as requiring that k_1 and k_2 be accumulated so that the incremental contributions of k_1 and k_2 to the objective functional be equal and that these contributions be equal to the marginal evaluation of additional consumption at every instant in time. Substituting (7.23) into the adjoint equations, we get

$$\dot{q}_1 = q_1 (\rho + \delta_1 - f' - g_1) \tag{7.24}$$

$$\dot{q}_2 = q_2 (\rho + \delta_2 + g_2)$$

which, together with (7.23), imply

$$f' - g_1 + g_2 = \delta_1 - \delta_2. \tag{7.25}$$

This interprets as requiring the difference in the marginal contribution of the optimum $k_1(t)$ to net output and the marginal contribution of the optimum $k_2(t)$ to damages reduction to be equal to the difference in depreciation rates.

In order to find out something about the state variables, differentiate (7.25) with respect to k_2 to get

$$(f'' - g_{11} + g_{21}) dk_1/dk_2 + g_{22} - g_{12} = 0 \tag{7.26}$$

which solves for dk_1/dk_2 as

$$dk_1/dk_2 = (g_{12} - g_{22})/(f'' - g_{11} + g_{21}). \tag{7.27}$$

In general it is difficult to place a sign restriction on (7.27), but in the special case that $g_{21} = g_{12} = 0$,[f] we can say that

$$dk_1/dk_2 < 0 \quad \text{when } |f''| > |g_{11}|, \text{ and} \tag{7.28}$$

$$dk_1/dk_2 > 0 \quad \text{when } |f''| < |g_{11}|.$$

The interpretation is that if diminishing returns to production sets in faster than diminishing returns to damages, abatement capital must be substituted for directly productive capital along the optimal path in order to maximize the objective. On the other hand, if diminishing returns sets in faster in damages

---

[f]This is the case when the marginal effectiveness of k_2 in reducing damages is independent of the output (k_1) level.

than in production, then both k_1 and k_2 are complementary and should be accumulated in some optimal proportion.

Boundary Solutions. In this section solutions for s_1 and s_2 which lie on the boundary of the control region are examined, and to do this remember from (7.22) that the partials of H with respect to s_i are

$$\partial H/\partial s_1 = (q_1 - q_2)s_2 y \tag{7.29}$$

$$\partial H/\partial s_2 = [(q_1 - q_2)s_1 + q_2 - U']y.$$

For $y > 0$, we can define,

$$H_i = \frac{1}{y}\frac{\partial H}{\partial s_i}, \quad i = 1,2. \tag{7.30}$$

Table 7-2 shows the gradient vector, $\nabla H = [H_1, H_2]$, corresponding to the various boundary policies, which is plotted to indicate what is happening to H at each of various boundary controls in Figure 7-2, Plate b. If the gradient of a

Table 7-2

Policy Number	Controls	$H = [H_1, H_2]$
1	$s_1 = s_2 = 1$	$H_1 = q_1 - q_2 \geq 0$
		$H_2 = q_1 - U'(0) \geq 0$
2	$s_1 = 1, s_2 = 0$	$H_1 = 0 \geq 0$
		$H_2 = q_1 - U'(y) \leq 0$
3	$s_1 = s_2 = 0$	$H_1 = 0 \leq 0$
		$H_2\, q_2 - U'(y) \leq 0$
4	$s_1 = 0, s_2 = 1$	$H_1 = q_1 - q_2 \leq 0$
		$H_2 = q_2 - U'(0) \geq 0$
5	$s_1 = 1, 0 \leq s_2 \leq 1$	$H_1 = q_1 - q_2 \geq 0$
		$H_2 = q_1 - U'[(1-s_2)y] = 0$
6	$s_1 = 0, 0 \leq s_2 \leq 1$	$H_1 = q_1 - q_2 < 0$
		$H_2 = q_2 - U'[(1-s_2)y] = 0$
7	$s_2 = 1, 0 \leq s_1 \leq 1$	$H_1 = 0$
		$H_2 = q_2 - U'(0) \leq 0$

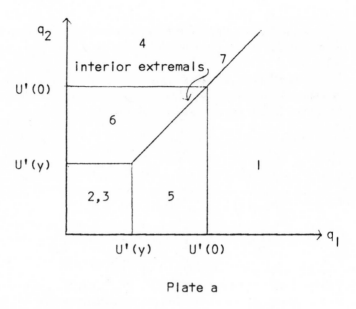

Plate a

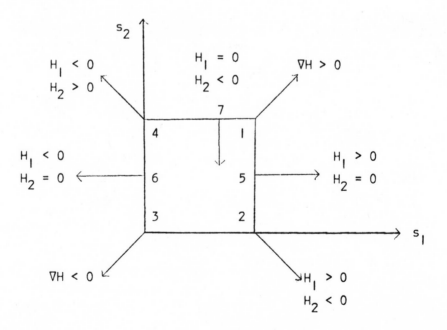

Plate b

Figure 7-2.

policy is such that a permissible move to the interior of the control region improves H, then that boundary policy is not an optimal one. Such a policy is number 7 ($s_2 = 1$, $0 \leqslant s_1 \leqslant 1$) as can be seen by the gradient which points to the interior of S. From Plate a in Figure 7-2, it can be seen that policies 1, 4, and 7 are not optimal policies in the case when $U'(0) = \infty$ as was assumed about the utility function. The marginal utility of consumption near zero is simply too high to warrant saving everything. This leaves policies 2 and 3, which are indistinguishable in their control effects, and policies 5 and 6 as possible optimal boundary policies to be used depending on the starting point of the system and its development over time.

Examination of this model demonstrates the nature of dynamic allocation problems. As soon as one permits damages due to pollution to be jointly produced with output, there is some optimal allocation of society's investment between abatement capital and directly productive capital (as pointed out, this may be a "boundary" allocation for some points in time). Moreover, since a decentralized market allocative mechanism no more achieves optimality in the dynamic context than in the static, the problem remains to find suitable alternative ways to provide proper allocative guidance to an economy with pollution externalities.

8 Conclusions

This work is an attempt to deal with the body of literature that has come to be known as the externality literature. Primarily, this consists of papers that have appeared in the past twelve or fifteen years in the journals with scattered references appearing before that time, and books, many of which have been published by Johns Hopkins for Resources for the Future.

The distinction between the pollution and nuisance models is clearly drawn. The primary distinction is that the pollution model treats residuals as an input into the consumption or production process, with the possibility of substitution (abatement), while the nuisance model affords no such substitution possibilities (abatement is impossible). In the former model the focus is on the optimal level of pollution which minimizes the sum of abatement costs and damage costs. The latter model focuses on the optimal level of consumption or production. The distinction may be obvious enough, but if it is, it is often confused or even ignored. In its most extreme form, this confusion takes the form of reaching policy conclusions of the following type: (1) limiting economic growth is the only way to limit environmental damage, (2) limiting industrial production is the only way to limit water pollution, (3) limiting the number of cars in an air basin is the only way to reduce air pollution, and (4) limiting agricultural production is the only way to limit the use of dangerous pesticides. These policies based on a faulty perception of which model to apply would cause environmental quality to cost much more than it otherwise would, and their advocacy may even slow down the environmental effort because of the drastic, though unwarranted and often untenable, policy implications. It is easy enough to see why the third world countries and the low-income people of the industrialized countries view even a well-conceived environmental effort as threatening, let alone one that reaches policy conclusions of the type presented above. It surely must seem like a cruel joke to the people of a developing country to hear a seriously maintained no-growth position in the name of the environment. To suggest, however, at some point that further growth in directly productive capital and consumption can be substituted for improvements in environmental quality is not nearly the bitter pill that the no-growth policy is.

Market structure was found to matter in the nuisance case but not in the pollution case. If there is a market structure of which it is true that the wrong[a] quantity is produced, it would hardly be appropriate to recommend a change in quantity because of a nuisance without regard to the misallocation due to the

_____
[a]Wrong in the Pareto sense.

139

market structure. The same is not true of the pollution case. Continuing the idea that pollution is an input, the misallocation that results from underpriced pollution has the same quality as the misallocation that results from underpriced labor. Too much of the input is used relative to other inputs regardless of the market structure in which the producer finds himself. If market structure is an effective constraint, then definition of the basinwide optimum in the use of pollution as an input with respect to the monopoly demand for pollution, and the subsequent achievement of that optimum, represents a Pareto improvement in the allocation of resources.

Voluntary mechanisms for the solution of the externality problem are frought with difficulties. Bargaining is least likely to occur because of prohibitive transactions costs in precisely the case where an optimal result could be expected, namely in the large numbers case. If there are only two parties to the bargain, nothing guarantees that the agreement reached is optimal. The fact that bargaining does not necessarily occur at the margin opens the possibility that a whole host of mutually beneficial but nonoptimal bargains could be struck on finite moves from the status quo. But this is precisely the case for which transactions costs are likely not to be prohibitive.

It is found that the placement of liability in a voluntary setting does not matter only under restrictive circumstances. If there are income effects, or even without income effects if utility or cost functions are nonseparable, the placement of liability affects the outcome. Moreover, since the placement of liability affects the distribution of rents and hence demand patterns and the long-run industry and output mix in response to the changed rent distribution and demand configuration, the long-run equilibrium must certainly depend on the placement of liability. To suppose that the world would look the same regardless of the placement of liability in anything but the most clinical of circumstances takes a great leap of faith.

Merger can resolve the externality problem nicely in a purely voluntary setting for the case of firms. But this leaves the case of firms and consumers or consumers and consumers unresolved. If there exists the possibility of involuntary transfers, then the incentive for firms to merge is greatly weakened. A successful suit enjoining the offending firm from its cost-imposing activity is always preferred to the merger solution by the offended firm. Yet from the social point of view the injunction may represent an excessively costly way of dealing with the externality.

Involuntary mechanisms include taxation, prohibition, and directives. It is found that appropriate taxes could be derived in both the pollution and nuisance case regardless of separability. The information costs of deriving precise Pareto taxes are high because information sufficient to solve the centralized planning problem is needed in general. An iterative procedure to approximate an optimal tax to any desired degree of accuracy in the nuisance case has been suggested, and the information requirements are almost entirely eliminated except for the recognition of the interdependence caused by the externality. In the pollution case the information required for setting optimal taxes is knowledge of the marginal abatement cost functions of the pollution sources and the marginal damage function. If there are more than a few pollution sources, information on the marginal abatement costs for each source would impose a substantial transaction cost. In addition, estimating damages as a function of ambient concentrations of pollutants is at an early stage, while estimates that are available often understate damages because of the exclusion of chronic and intangible, as opposed to acute, effects.

Arguments are developed for the case that a system of payments is not fully equivalent to a system of charges. The first argument centers on the fact that there is no natural origin from which payments should be made, and since there is no natural origin, the perverse incentives of a payments scheme induce polluters to make the origin larger than otherwise. The inducement to pollute more (make the origin larger) continues even if costs are incurred so long as the marginal cost of additional pollution is less than the marginal payment from later "abating" that incremental pollution. There are also differential long-run effects depending on whether there are payments or charges. The distribution of rents and hence demand patterns, and industry and output mix are different depending on whether a payments or charges scheme is used. A third consideration is that payments must be made to those who leave the basin or potential entrants into a basin who would enter but for the payments. Payments must continue to firms who find it more profitable to shut down as well. This makes the administration of an effective payments scheme much more difficult because the administrative agency must keep track not only of the current polluters in the basin, but the potential polluters, the polluters that have left the basin, and the polluters that have shut down. In a charges scheme only the current polluters are dealt with.

This leads to the final critique of a payments scheme and this has to do with people's sense of equity. By convention the economy of private goods and services has evolved into a system where charges are levied for the use of scarce resources rather than payments being made for the avoidance of scarce resources. The extension of the convention to the case of externalities is natural enough and rests on the same underlying sense of equity that supports the convention in the economy of private goods.

In both the pollution and nuisance cases, a tax scheme functions optimally with a tax being placed only on the offender. No compensation to or tax on the offended party is necessary for optimality to be achieved.

There is one case where an externality problem does not yield to a taxation solution, and that is the case of a Goetz-Buchanan industry. If each firm in the industry experiences a nuisance which is a function of the industry output produced externally to the firm, but the output does not cause any external effects to either consumers or firms in other industries, then the external effects within the industry can be completely eliminated by having a monopoly. Since there is no industry output being produced externally to the monopolist, the Goetz-Buchanan industry is one in which there are external economies of scale. Just as internal economies of scale cannot be captured by taxation, the external economies available through having fewer but larger producers in the industry cannot be captured by placing a unit tax on the output of each firm.

The conditions that must hold for a prohibition to be optimal from the Pareto point of view are obvious enough. Directives which include standards and requirements are possibilities for controlling externalities. Most of the problems with directives center on incentives and information requirements. The incentives in the case of directives are upside down from what they are in the case of taxation. There is never any incentive to do any better than the standard dictates or to do any more than the requirement stipulates. Moreover there is every incentive to undermaintain, cheat, and chisel if the expected value of the sanctions, given enforcement, is less than the cost of complying. Finally, to set optimal point-by-point standards or to stipulate optimal requirements necessitates all the information necessary to a centralized planning problem. This is likely to be prohibitively expensive with the result that standards or requirements are nonoptimal.

It is found that the separate facilities approach can be expected to lead to optimal results in a great many cases of nuisance as, for example, in mediating the conflicting interests of smokers and nonsmokers, or of swimmers and surfers. But it probably is not the answer in mediating the interests of drivers and pure-air enthusiasts.

In approaching the outstanding problems of air and water pollution a combination of policies is advocated by some. Dales suggests setting more or less arbitrary ambient standards for water quality and then issuing pollution rights in an amount that would achieve this standard. These rights are to be auctioned off and no one could pollute without holding a right to do so. In a similar vein, Baumol proposes setting ambient standards and raising taxes to the point where the standard is met. In both cases whatever the standard, it is achieved at a minimum of cost. The *Genossenschaften* of the Ruhr Valley use a combination of charges, prohibitions, and separate facilities in order to deal with waterborne waste management in West Germany. The associations of the Ruhr also operate treatment plants to take advantage of any economies of scale or locational advantages. This approach is the one recommended by Roberts for the attack on water pollution in this country, and it extends naturally to problems of air pollution. There are at least two points worth mentioning about this approach. First, charges play an important part in getting polluters to realize the costs of their actions and to economize on a scarce input, residuals handling. Charges may also be expected to provide the incentive for intrafirm as well as interfirm optimality. Second, the authority is a basinwide authority which corresponds to the problem shed.

Finally, the necessity of reliance on an "agency" in the case of involuntary mechanisms means perforce that the political process must be relied upon. Buchanan points out that if people's motivations in the political process do not differ from their motivations in the market setting, that there is no reason to suspect that the political process can do any better than the market mechanism.[1] Policy may very well cause marginal conditions to be violated at different margins and for different individuals. That is policy creates externalities of its own. This means that the choice is probably not between optimality and suboptimality, the real choice is between various kinds and degrees of suboptima. The appeal of the Dales-Baumol approach is that political reform is divorced from the problems of efficiency in environmental resource use. Whether the ambient standard is set optimally by an omniscient agency, or suboptimally by a corrupt political process, whatever standard is set and achieved in the Dales-Baumol approach is achieved at a minimum of cost. If the standard is too lenient due to political corruption, but the standard is achieved efficiently, then there is no room for the argument from those involved in the intrigue that there could be a lot more environmental quality for the same cost if only there were efficiency in abatement. The problem focuses squarely on the nature of the political process which sets the lenient standard and its reform, rather than the issue of how a given standard is achieved. The problems are separable in that sense.

But this discussion points out that the benevolent despot theory of "agency" behavior is probably a bad model. This is where questions of representation (what interests are represented on the agency?), accountability (to the voters directly, or to an elected official?), and financing practices (if agency representatives are elected) enter into the problem. There are a whole host of legal and political issues dealing with altering the political weights of various interest groups, providing starting points for the mediation of conflicting interests, and in general providing a legal framework within which to operate (are agency bureaucrats free to enforce at their discretion?). These are interesting and important questions on which lawyers, economists, and political theorists are working. Hopefully, however, this book has helped to clarify issues surrounding the efficiency question of policies designed to deal with externalities.

Notes

Notes

Note to Preface

1. See T.S. Kuhn, *The Structure of Scientific Revolutions*, 2d ed. (Chicago: University of Chicago Press, 1970).

Notes to Chapter 1
Introduction

1. Staff, Committee on Public Works, United States Senate, "A Study of Pollution—Air," September 1963, in H. Wolozin (ed.), *The Economics of Air Pollution* (New York: W.W. Norton, 1966), p. 205.

2. L.B. Lave, "Air Pollution Damage: Some Difficulties in Estimating the Value of Abatement," in A.V. Kneese and B.T. Bower (eds.) *Environmental Quality Analysis* (Baltimore: Johns Hopkins Press for Resources for the Future, 1972), p. 240.

3. See The President's Council on Environmental Quality, *Environmental Quality*, Annual Reports, (Washington, D.C.: U.S. Government Printing Office, Beginning 1970).

4. M.I. Goldman, "The Convergence of Environmental Disruption," *Science* 170 (October 2, 1970): 37-42; also M.I. Goldman, "Growth and Environmental Problems of Noncapitalist Nations," *Challenge* 16 (July/August 1973): 45-51.

5. For distributional considerations, see for example, A.M. Freeman, III, "Distribution of Environmental Quality," in A.V. Kneese and B.T. Bower (eds.), *Environmental Quality Analysis* (Baltimore: Johns Hopkins Press for Resources for the Future, 1972): 243-78; M.S. Feldstein, "Distributional Equity and the Optimal Structure of Public Prices," *American Economic Review* 62 (1972): 32-36.

6. See P.D. Bush, "The Normative Implications of Positive Analysis," Western Economic Association Meetings, August 1968, (mimeo).

Notes to Chapter 2
The Theory of Market Behavior

1. The basic reference for the proof of this theorem is K.J. Arrow, "An Extension of the Basic Theorems of Classical Welfare Economics," in J. Neyman (ed.) *Proceedings of the Second Berkeley Symposium on Mathematical Statistics and Probability* (Berkeley: University of California Press, 1951), pp. 507-32. See also F.M. Bator, "The Simple Analytics of Welfare Maximization," *American Economic Review* 47 (March 1957): 22-59.

2. K. Arrow, op. cit., p. 527.

3. Ibid., p. 511.

4. H. Kuhn and A. Tucker, "Nonlinear Programming," in J. Neyman (ed.) op. cit., pp. 481-92; see also S. Karlin, *Mathematical Methods and Theory in Games, Programming, and Economics*, Vol. I (Reading, Massachusetts: Addison-Wesley, 1959), pp. 216-17.

5. T. Negishi, "Welfare Economics and Existence of an Equilibrium for a Competitive Economy," *Metroeconomica* 11 (1960): 92-97.

6. S. Karlin, op. cit., p. 201. Much of this discussion of nonlinear programming follows Karlin, op. cit., Chapter 7, but the primary source is H. Kuhn and A. Tucker, op. cit.

7. K. Arrow, op. cit., p. 507.

Notes to Chapter 3
The Theory of External Effects: Nuisance

1. For a discussion of this debate, see H.S. Ellis and W. Fellner, "External Economies and Diseconomies," *American Economic Review* 33 (September 1943), pp. 493-511, and E.J. Mishan, "The Post War Literature on Externalities: An Interpretive Essay," *Journal of Economic Literature* 9, (March 1971): 1-28.

2. A.C. Pigou, *The Economics of Welfare*, 2d ed. (London: Macmillan and Co., 1924), p. 152. Reprinted by permission of Macmillan, London and Basingstoke.

3. Ibid., p. 161.

4. "Cost Curves and Supply Curves," *Zeitschrift für Nationalökonomie* 3 (September 1931): 23-46. See also T. Scitovsky, "Two Concepts of External Economies," *Journal of Political Economy* 62 (1954): 143-51, for further elaboration. For an excellent treatment of pecuniary externalities, or what he calls price effects, see P. Bohm, *External Economies in Production* (Stockholm: Almquist and Wicksell, 1964).

5. J.E. Meade, "External Economies and Diseconomies in a Competitive Situation," *The Economic Journal* 62 (1952): 54-67.

6. J.M. Buchanan and W.C. Stubblebine, "Externality," *Economica* 29 (November 1962): 371-84. See also T. Scitovsky, op. cit., p. 144, for categorizing the various technological interdependencies that are possible, i.e., interdependencies among consumers, among producers, and among producers and consumers.

7. Buchanan and Stubblebine, loc. cit.

8. Ibid.

9. H.S. Ellis and W. Fellner, "External Economies and Diseconomies," *American Economic Review* XXXIII (September 1943), p. 511.

10. F.M. Bator, "The Anatomy of Market Failure," *Quarterly Journal of Economics* 72 (August 1958): 351-79.

11. J.G. Head, "Public Goods and Public Policy," *Public Finance*, 17 (1962): 197-221.

12. E.J. Mishan, "Reflections on Recent Developments in the Concept of External Effects," *Canadian Journal of Economics and Political Science* 31 (February 1965): 6.

13. Bator, op. cit.

14. P.A. Samuelson, "The Pure Theory of Public Expenditure," *Review of Economics and Statistics* 36 (November 1954): 387-89.

15. J.M. Buchanan, "Joint Supply, Externality and Optimality," *Economica* 33 (1966): 404-15.

16. Buchanan and Stubblebine, op. cit.

17. K.J. Arrow, "The Organization of Economic Activity: Issues Pertinent to the Choice of Market versus Nonmarket Allocation," in Joint Economic Committee of Congress, *The Analysis and Evaluation of Public Expenditures: The P.P.B. System* (Washington: U.S. Government Printing Office, 1969), pp. 47-66.

18. E.J. Mishan, "The Relationship between Joint Products, Collective Goods, and External Effects," *Journal of Political Economy* 77 (1969): 343.

19. Ibid.

20. P.A. Samuelson, "Contrast between Welfare Conditions for Joint Supply and for Public Goods," *Review of Economics and Statistics* 51 (1969): 26-30. See also A.W. Evans, "Private Good, Externality, and Public Good," *Scottish Journal of Political Economy* 17 (February 1970): 79-89.

21. J.M. Buchanan, "External Diseconomies, Corrective Taxes, and Market Structure," *American Economic Review* 59 (March 1969): 175.

22. F.T. Dolbear, "On the Theory of Optimum Externality," *American Economic Review* 57 (1967): 91-103.

23. Buchanan and Stubblebine, loc. cit.

24. H. Kuhn and A. Tucker, "Nonlinear Programming," in J. Neyman (ed.) *Proceedings of the Second Berkeley Symposium on Mathematical Statistics and Probability* (Berkeley: University of California Press, 1951), pp. 481-92; see also Samuel Karlin, op. cit., pp. 216-17.

25. T. Negishi, "Welfare Economics and Existence of an Equilibrium for a Competitive Economy," *Metroeconomica* 11 (1960): 92-97.

26. W.J. Baumol, "External Economies and Second-Order Optimality Conditions," *American Economic Review* 54 (June 1964): 358-72. See especially Theorem 2, p. 364.

27. J.M. Buchanan, loc. cit.

28. See for example, O. Davis and A. Whinston, "Externalities, Welfare, and the Theory of Games," *Journal of Political Economy* 70 (1962): 241-62, and S. Wellisz, "On External Diseconomies and the Government Assisted Invisible Hand," *Economica* 31 (n.s. 1964): 345-62.

29. See for example, A.V. Kneese and B.T. Bower, *Managing Water Quality: Economics, Technology, Institutions* (Baltimore: Johns Hopkins University Press for Resources for the Future, 1968), pp. 94-96 and pp. 110-16.

30. See for example, Davis and Whinston [1962], op. cit.; O. Davis and A. Whinston, "On Externalities, Information, and the Government Assisted Invisible Hand," *Economica* 33 (n.s. 1966): 303-18; and S. Wellisz, op. cit.

31. For an example of this confusion, see an otherwise excellent treatment by J. Burkhead and J. Miner, *Public Expenditure* (Chicago: Aldine, 1971), p. 115, note 5.

32. J.M. Buchanan and M. Kafoglis, "A Note on Public Goods Supply," *American Economic Review* 53 (June 1963): 403-14.

33. P.E. Vincent, "Reciprocal Externalities and Optimal Input and Output Levels," *American Economic Review* 59 (1969): 976-83.

34. Ibid., p. 976.

35. W.J. Baumol, "External Economies and Second-Order Optimality Conditions," *American Economic Review* 54 (June 1964): 358-72.

36. L.D. Schall, "Technological Externalities and Resource Allocation," *Journal of Political Economy* 79 (September-October 1971): 983-1001.

37. See also M. Olson, Jr., and R. Zeckhauser, "The Efficient Production of External Economies," *American Economic Review* 60 (June 1970): 512-17.

38. P.A. Diamond and J.A. Mirrlees, "Aggregate Production with Consumption Externalities," *Quarterly Journal of Economics* 87 (February 1973): 1-24.

39. Ibid., p. 12.

40. Ibid., p. 7.

41. Ibid., p. 12.

42. Ibid., p. 20.

43. Buchanan, "External Diseconomies," pp. 174-77.

44. Arrow, "Organization of Economic Activity," pp. 56-9.

45. H. Demsetz, "The Exchange and Enforcement of Property Rights," *Journal of Law and Economics* 7 (October 1964): 11-26.

Notes to Chapter 4
The Theory of External Effects:
Pollution and Congestion

1. J. Rothenberg, "The Economics of Congestion and Pollution: An Integrated View," *American Economic Review* 60 (1970): 114-21; see especially, p. 115.

2. This point is emphasized by R.U. Ayres, A.V. Kneese, "Production, Consumption, and Externalities," *American Economic Review* 59 (June 1969): 284, who cite F. Knight and I. Fisher.

3. D.A. Starrett, "Fundamental Nonconvexities in the Theory of Externalities," *Journal of Economic Theory* 4 (1972): 190.

4. Ibid., p. 192, Proposition 4.

5. H.R. Bowen, *Toward Social Economy* (New York: Rinehart, 1948), p.

177; P.A. Samuelson, "The Pure Theory of Public Expenditure," *The Review of Economics and Statistics* 36 (November 1954): 387-89; P.A. Samuelson, "Diagrammatic Exposition of a Theory of Public Expenditure," *Review of Economics and Statistics* 37 (November 1955): 350-56.

6. A.V. Kneese, "Air Pollution–General Background and Some Economic Aspects," in H. Wolozin (ed.) *The Economics of Air Pollution* (New York: W.W. Norton, 1966), pp. 33-34.

7. Ibid.

8. R. Turvey, "On Divergences between Social Cost and Private Cost," *Economica* 30 (August 1963): 309-13.

9. T.D. Crocker, "The Structuring of Atmospheric Pollution Control Systems," in H. Wolozin (ed.) *The Economics of Air Pollution* (New York: W.W. Norton, 1966), p. 69.

10. W.J. Baumol, "On Taxation and the Control of Externalities," *American Economic Review* 62 (June 1972): 319, note 15.

11. R.G. Lipsey and K. Lancaster, "The General Theory of Second Best," *Review of Economic Studies* 24 (February 1956): 11-32.

12. C.C. Morrison, "Generalizations on the Methodology of the Second Best," *Western Economic Journal* 6 (1968): 120.

13. Ibid.; see also O.A. Davis and A.B. Whinston, "Piecemeal Policy in the Theory of Second Best," *Review of Economic Studies* 34 (July 1967): 330ff; and E.J. Mishan, "Second Thoughts on Second Best," *Oxford Economic Papers* 14 (October 1962): 205-17.

Notes to Chapter 5
Policy to Achieve Optimality: Nuisance

1. R. Coase, "The Problem of Social Cost," *Journal of Law and Economics* 3 (October 1960): 1-44.

2. Closely paraphrased from Coase, op. cit., p. 6.

3. E.J. Mishan, "Pareto Optimality and the Law," *Oxford Economic Papers* 19 (1967): 255-87; E.J. Mishan, "Pangloss on Pollution," *Swedish Journal of Economics* 73 (March 1971): reprinted in P. Bohm and A.V. Kneese, *The Economics of Environment* (London: Macmillan, 1971), pp. 66-73; F.T. Dolbear, "On the Theory of Optimum Externality," *American Economic Review* 57 (1967): 90-103; P. Burrows, "On External Costs and the Visible Arm of the Law," *Oxford Economic Papers* 22 (1970): 39-55.

4. Dolbear, op. cit., p. 91.

5. Ibid. At this point, Dolbear also makes a reference to Buchanan and Stubblebine's fence problem, which further emphasizes that the externality he is considering is of the nonabatable type.

6. See also E.J. Mishan, "Welfare Criteria for External Effects," *American Economic Review* 51 (1961): especially p. 603.

7. P. Burrows, op. cit.; see also E. Furubotn and S. Pejovich, "Property Rights and Economic Theory: A Survey of Recent Literature," *Journal of Economic Literature* 10 (December 1972): 1143, and T.E. Borcherding, "Liability in Law and Economics: Note," *American Economic Review* 60 (December 1970): 946-48, who also recognize that the Coase Theorem abstracts from income effects.

8. J.R. Marchand and K.P. Russell, "Externalities, Liability, Separability, and Resource Allocation," *American Economic Review* 63 (September 1973): 611-20.

9. A. Gifford, Jr., and C.C. Stone, "Externalities, Liability and the Coase Theorem: A Mathematical Analysis," *Western Economic Journal* 11 (September 1973): 260-69, defend Coase abstracting income effects, but they ignore the separability issue which proves to make a difference.

10. G. Calabresi, "The Decision for Accidents: An Approach to Nonfault Allocation of Costs," *Harvard Law Review* 78 (1965): 713-30. G.W. Nutter, "Coase Theorem on Social Cost: A Footnote," *Journal of Law and Economics* 10-11 (1967-68): 503-507, defends Coase against Calabresi. The defense is largely a restatement of Coase. Calabresi uses the word "rents" because were it not for Ricardian rents, a competitive firm would have nothing with which to bargain, there being no economic profit. On this, see S. Wellisz, "On External Diseconomies and the Government Assisted Invisible Hand," *Economica* 31 (1964): 351.

11. G. Calabresi, "Transaction Costs, Resource Allocation, and Liability Rules," *Journal of Law and Economics* 11 (April 1968): 67-73.

12. See also D.A. Starrett, op. cit., p. 190.

13. Calabresi, op. cit., p. 68.

14. Ibid.

15. R.O. Zerbe, "Theoretical Efficiency in Pollution Control," *Western Economic Journal* 8 (December 1970): 368.

16. Wellisz, "On External Diseconomies and the Government Assisted Invisible Hand," *Economica* 31 (1964): 352.

17. Ibid., p. 352, note 1.

18. K.J. Arrow, "Political and Economic Evaluation of Social Effects and Externalities," in M.D. Intriligator (ed.) *Frontiers of Quantitative Economics* (Amsterdam: North Holland, 1971), p. 17. This is the same paper that is given under the title, "The Organization of Economic Activity: Issues Pertinent to the Choice of Market versus Nonmarket Allocation," in Joint Economic Committee of Congress, *The Analysis and Evaluation of Public Expenditures: The P.P.B. System* (Washington: U.S. Government Printing Office, 1969), pp. 47-66.

19. Wellisz, op. cit., p. 353.

20. D.C. Shoup, "Theoretical Efficiency in Pollution Control: Comment," *Western Economic Journal* 9 (September 1971): 310-13.

21. Marchand and Russell, op. cit., p. 617.

153

22. G.A. Mumey, "The 'Coase Theorem': A Reexamination," *Quarterly Journal of Economics* 85 (1971): 721.

23. Ibid., p. 722.

24. R.O. Zerbe, "Theoretical Efficiency in Pollution Control: Reply," *Western Economic Journal* 9 (September 1971): 316.

25. H. Demsetz, "Theoretical Efficiency in Pollution Control: Comments on Comments," *Western Economic Journal* 9 (December 1971): 444-46.

26. Zerbe, "Reply," op. cit., p. 316; and Demsetz, op. cit., pp. 445-46.

27. Demsetz, loc. cit.

28. Mumey, op. cit., pp. 722-23.

29. Demsetz, loc. cit.

30. H. Demsetz, "The Exchange and Enforcement of Property Rights," *Journal of Law and Economics* 7 (1964): 11-26.

31. J.M. Buchanan, *The Demand and Supply of Public Goods* (Chicago: Rand McNally, 1968), Chapter 5; see also G. Hardin, "The Tragedy of the Commons," *Science* 162 (1968): 2143-48 and T.C. Schelling, "On the Ecology of Micromotives," *The Public Interest* 25 (1971): 61-91. These writers discuss the free-rider problem in other contexts, but the common element is that a decision which is rational from the individual viewpoint is not rational from the group, or common viewpoint.

32. O. Davis and A. Whinston, "Externalities, Welfare and the Theory of Games," *Journal of Political Economy* 70 (1962): 241-62.

33. Ibid., p. 244.

34. R. Coase, loc. cit.

35. A.C. Pigou, *The Economics of Welfare*, 2d ed. (London: Macmillan and Co., 1924).

36. O. Davis and A. Whinston, loc. cit.

37. Davis and Whinston, op. cit., p. 255.

38. J.M. Buchanan, "External Diseconomies, Corrective Taxes, and Market Structure," *American Economic Review* 59 (March 1969): 174-77.

39. J.M. Buchanan and W.C. Stubblebine, "Externality," *Economica* 29 (November 1962): 371-84.

40. R. Turvey, "On Divergences between Social Cost and Private Cost," *Economica* 30 (August 1963): 309-13.

41. F.T. Dolbear, loc. cit.

42. S. Wellisz, op. cit., p. 359.

43. P.A. Diamond, "Consumption Externalities and Imperfect Corrective Pricing," *Bell Journal of Economics and Management Science* 4 (Autumn 1973): 526-38. The tax derivation follows Diamond closely.

44. Diamond, op. cit., p. 530.

45. W.J. Baumol, "On Taxation and the Control of Externalities," *American Economic Review* 62 (June 1972): 307-22.

46. Baumol, op. cit., p. 307.

47. C.J. Goetz and J.M. Buchanan, "External Diseconomies in Competitive Supply," *American Economic Review* 61 (1971): 883-90.

48. The author has simplified the classification of administrative solutions offered by O.A. Davis and M.I. Kamien, "Externalities, Information, and Alternative Collective Action," in Joint Economic Committee, United States Congress, *The Analysis and Evaluation of Public Expenditures: The PPB System*, Vol. I (Washington, D.C.: U.S. Government Printing Office, 1969), pp. 67-86.

49. E.J. Mishan, *The Costs of Economic Growth* (New York: Frederick Praeger, 1967), Chap. viii.

50. H.A. Thomas, Jr., "The Animal Farm: A Mathematical Model for the Discussion of Social Standards for the Control of the Environment," *Quarterly Journal of Economics*, (February 1963), reprinted in R. Dorfman and N.S. Dorfman (eds.) *Economics of the Environment: Selected Readings* (New York: W.W. Norton, 1972), pp. 250-56.

51. Ibid., p. 255.

52. See P. Diamond, "Consumption Externalities," op. cit.

53. O. Davis and A.B. Whinston, "On Externalities, Information, and the Government Assisted Invisible Hand," *Economica* 33 (1966): 310.

54. Ibid., p. 310.

55. Ibid., pp. 312-18.

56. D.A. Starrett, "Fundamental Nonconvexities in the Theory of Externalities," *Journal of Economic Theory* 4 (1972): 195.

Notes to Chapter 6
Policy to Achieve Optimality:
Pollution and Congestion

1. Marchand and Russell, loc. cit.

2. C.R. Plott, "Externalities and Corrective Taxes," *Economica* 33 (February 1966): 84-87, notes that placing the charge on output is incorrect except under certain circumstances, those being when the externality is a nuisance. In the pollution case Plott emphasizes that the charge must be placed on the ". . . smoke, or, under certain conditions, on the resource input from which the smoke is generated." (p. 84). See also R. Turvey, "On Divergences between Social Cost and Private Cost," *Economica* 30 (August 1963): 309-13, who emphasizes that the tax must be levied in the amount of the incremental damage (cf. p. 312).

3. R. Coase, op. cit., pp. 41-42.

4. See also Buchanan and Stubblebine, op. cit., Section III, who argue that it is necessary for optimality that a double tax system be instituted.

5. W.J. Baumol, "On Taxation and the Control of Externalities," *American Economic Review* 62 (June 1972): 312, note 8.

6. Ibid., pp. 309-12.

7. Ibid., p. 311. As noted earlier, bargaining between the parties must not be economically feasible for the tax solution to result in optimality. Otherwise with a tax the two parties have the incentive to reach a bargaining solution which departs from the optimal one.

8. Ibid., p. 312.

9. Ibid., pp. 312-13. Baumol indicates that the externality itself prevents too many residents. As long as residents are rational, no tax on them is needed.

10. M.J. Kamien, N.L. Schwartz, and F.T. Dolbear, Jr., "Asymmetry between Bribes and Charges," *Water Resources Research* 2 (First Quarter, 1966): pp. 147-57; R.M. Solow, "The Economist's Approach to Pollution and Its Control," *Science* 173 (July-September 1971): 498-503; E.S. Mills, "Economic Incentives in Air-Pollution Control," in H. Wolozin (ed.) *The Economics of Air Pollution* (New York: W.W. Norton, 1966), p. 45.

11. M.J. Kamien, N.L. Schwartz, and F.T. Dolbear, Jr., loc. cit.

12. A.M. Freeman, III, "Bribes and Charges: Some Comments," *Water Resources Research* 3 (Fall Quarter, 1967): 288. Copyrighted by American Geophysical Union.

13. D.F. Bramhall and E.S. Mills, "A Note on the Symmetry between Fees and Payments," *Water Resources Research* 2 (Third Quarter, 1966): 615-16.

14. A.V. Kneese, *The Economics of Regional Water Quality Management* (Baltimore: Johns Hopkins Press, 1964), p. 196; E.S. Mills, op. cit.; R.M. Solow, op. cit.

15. A.V. Kneese, op. cit., p. 196.

16. A.V. Kneese, *Managing Water Quality: Economics, Technology and Institutions* (Baltimore: Johns Hopkins Press for Resources for the Future, 1968), p. 51.

17. References for the idea that requirements are unlikely to achieve the interfirm and intrafirm least cost property include among others H. Wolozin, "The Economics of Air Pollution Control: Central Problems," *Law and Contemporary Problems* 33 (1968): 229-33; and R.M. Solow, loc. cit.

18. H. Wolozin, loc. cit.

19. J.H. Dales, *Pollution, Property and Prices* (Toronto: University of Toronto Press, 1968), p. 84ff.

20. A.V. Kneese and B.T. Bower, op. cit., pp. 137-38.

21. H.W. Streeter and E.B. Phelps, *A Study of the Pollution and Natural Purification of the Ohio River—III: Factors Concerned in the Phenomena of Oxydation and Recreation*, Public Health Bulletin Number 146 (Washington, D.C.: U.S. Government Printing Office, February 1925); see also A.A. Teller, "The Use of Linear Programming to Estimate the Cost of Some Alternative Air Pollution Abatement Policies," *Proceedings*, IBM Scientific Computing Symposium on Water and Air Resources Management, October 23-25, 1967, Yorktown Heights, New York.

22. W.J. Baumol, "On Taxation," op. cit., p. 318ff.

23. W.J. Baumol and W.E. Oates, "The Use of Standards and Prices for the Protection of the Environment," in P. Bohm, and A.V. Kneese (eds.) *The Economics of Environment* (London: Macmillan, 1971), pp. 53-65.

24. J.H. Dales, op. cit., p. 33ff; also J.H. Dales, "Land, Water, and Ownership," *Canadian Journal of Economics* 34 (November 1968).

25. M.J. Roberts, "Organizing Water Pollution Control: The Scope and Structure of River Basin Authorities," *Public Policy* 19 (Winter 1971): 75-141; also M.J. Roberts, "River Basin Authorities: A National Solution to Water Pollution," *Harvard Law Review* 83 (1970): 1527-56.

26. A.V. Kneese, *The Economics of Regional Water Quality Management* (Baltimore: Johns Hopkins Press for Resources for the Future, 1964), p. 200ff.

27. Ibid., p. 145ff.

28. A.V. Kneese and B.T. Bower, op. cit., Chapter 12, for a description and critique of the *Genossenschaften*.

29. Ibid., p. 173; M.J. Roberts, "Organizing. . . ," op. cit., p. 141.

30. A.V. Kneese and B.T. Bower, op. cit., pp. 136-37.

Notes to Chapter 7
Theory of External Effects: General
Equilibrium and Dynamic Optimality

1. W. Leontief, "Environmental Repercussions and the Economic Structure: An Input-Output Approach," *Review of Economics and Statistics* 52 (August 1970): 374-81.

2. R.U. Ayres and A.V. Kneese, "Production, Consumption and Externalities," *American Economic Review* 59 (June 1969): 282-97; A.V. Kneese and R.C. D'Arge, "Pervasive External Costs and the Response of Society," 91st U.S. Congress, Joint Economic Committee, Subcommittee on Economy in Government, *The Analysis and Evaluation of Public Expenditures: The P.P.B. System* (Washington: U.S. Government Printing Office, 1969); A.V. Kneese, R.U. Ayres, and R.C. D'Arge, *Economics and the Environment* (Baltimore: Johns Hopkins Press for Resources for the Future, 1970). See P.A. Victor, *Pollution: Economy and Environment* (London: George Allen and Unwin, Ltd., 1972), Chapter 2, for a thorough review and critique of this literature.

3. C.S. Russell, and W.O. Spofford, "A Quantitative Framework for Residuals Management Decisions," in A.V. Kneese and B.T. Bower, (eds.) *Environmental Quality Analysis* (Baltimore: Johns Hopkins Press for Resources for the Future, 1972), Chapter 4. See also A.V. Kneese, "Environmental Pollution: Economics and Policy," *American Economic Review* 61 (May 1971): 153-66; C.S. Russell, "Application of Microeconomic Models to Regional Environmental Quality Management," *American Economic Review* 63 (May 1973): 236-43; and

C.S. Russell, "Models for Investigation of Industrial Response to Residuals Management Actions," in P. Bohm and A.V. Kneese (eds.) *The Economics of Environment* (London: Macmillan, 1971), pp. 141-63.

4. R. Dorfman, "Discussion," *American Economic Review* 63 (May 1973): 253-56.

5. Ibid.

6. P. Bohm, *External Economies in Production* (Stockholm: Almquist and Wiksell, 1964), p. 56.

7. See for example, E. Keeler, M. Spence and R. Zeckhauser, "The Optimal Control of Pollution," *Journal of Economic Theory* 4 (1971): 19-34; and V.L. Smith, "Dynamics of Waste Accumulation: Disposal versus Recycling," *Quarterly Journal of Economics* 86 (1972): 600-616.

8. L.S. Pontryagin, V.G. Boltyanskii, R.V. Gamkrelidze, and E.F. Meshchenko, *The Mathematical Theory of Optimal Processes* (New York: Interscience Publishers, 1962). The Maximum Principle bears approximately the same relationship to classical variational techniques that Kuhn-Tucker theory bears to the classical calculus.

9. F. Hahn, "Equilibrium Dynamics with Heterogeneous Capital Goods," *Quarterly Journal of Economics* 80 (1966): 633-46.

10. P.A. Samuelson, "Indeterminacy of Development in a Heterogeneous Capital Model with Constant Saving Propensity," in K. Shell (ed.) *Essays on the Theory of Optimal Economic Growth* (Cambridge: M.I.T. Press, 1967), Essay XII.

11. P.A. Samuelson and R.M. Solow, "A Complete Capital Model Involving Heterogeneous Capital Goods," *Quarterly Journal of Economics* 70 (1956): 537-62.

12. F.P. Ramsey, "A Mathematical Theory of Saving," *Economic Journal* 38 (1928): 543-49.

13. See, for example, J. Pitchford, "Population and Optimal Growth," *Econometrica* 40 (January 1972): 109-136.

Note to Chapter 8
Conclusions

1. J.M. Buchanan, "Politics, Policy, and Pigouvian Margins," *Economica* 29 (February 1962): 17-28.

Bibliography

Bibliography

Arrow, K.J. "An Extension of the Basic Theorems of Classical Welfare Economics." *Proceedings of the Second Berkeley Symposium on Mathematical Statistics and Probability*. Edited by J. Neyman. Berkeley: University of California Press, 1951.

_____. "The Organization of Economic Activity: Issues Pertinent to the Choice of Market versus Nonmarket Allocation." Joint Economic Committee of Congress. *The Analysis and Explanation of Public Expenditures: The P.P.B. System*. Washington: U.S. Government Printing Office, 1969.

_____. "Political and Economic Evaluation of Social Effects and Externalities." *Frontiers of Quantitative Economics*. Edited by M.D. Intriligator. Amsterdam: North Holland, 1971.

_____. *Social Choice and Individual Values*. Cowles Commission Monograph No. 12. New York: John Wiley and Sons, 1951.

_____, and Kurz, M. "Optimal Growth with Irreversible Investment in a Ramsey Model." *Econometrica* 38 (March 1970): 331-44.

Ayer, J. "Water Quality Control at Lake Tahoe: Dissertation on Grasshopper Soup." *California Law Review* 58 (November 1970): 1273-1330.

Ayres, R.U. and Kneese, A.V. "Pollution and Environmental Quality." *The Quality of the Urban Environment*. Edited by Harvey S. Perloff. Baltimore: Johns Hopkins Press for Resources for the Future, 1968.

_____, and Kneese, A.V. "Production, Consumption and Externalities." *American Economic Review* 59 (June (1969): 282-97.

_____, and McKenna, R.P. *Alternatives to the Internal Combustion Engine: Impacts on Environmental Quality*. Baltimore: Johns Hopkins Press for Resources for the Future, 1971.

Bailey, M.C. "Analytical Framework for Measuring Social Costs." *Journal of Farm Economics* 44 (May 1962): 564-74.

Bain, J.S. "Criteria for Undertaking Water-Resource Developments." *American Economic Review* 50 (March-May 1960): 310-20.

Bator, F.M. "The Anatomy of Market Failure." *The Quarterly Journal of Economics* 72 (August 1958): 351-79.

_____. "The Simple Analytics of Welfare Maximization." *American Economic Review* 47 (March 1957): 22-59.

Baumol, W.J. "Activity Analysis." *American Economic Review* 48 (June-December 1958): 837-73.

_____. "External Economies and Second-Order Optimality Conditions." *American Economic Review* 54 (June 1964): 358-72.

_____. "On Taxation and the Control of Externalities." *American Economic Review* 62 (June 1972): 307-22.

_____, and Oates, W.E. "The Use of Standards and Prices to Protect the

Environment." *The Swedish Journal of Economics* 73 (March 1971): 42-52. Reprinted in *The Economics of Environment.* Edited by P. Bohm and A.V. Kneese. London: Macmillan, 1971.

Becker, G.S. "A Theory of the Allocation of Time." *Economic Journal* 75 (September 1965): 493-517.

Bohm, P. "An Approach to the Problem of Estimating Demand for Public Goods." *The Economics of Environment.* Edited by P. Bohm and A.V. Kneese. London: Macmillan, 1971.

_____. *External Economies in Production.* Stockholm: Almquist and Wiksells, 1964.

_____, and Kneese, A.V., eds. The Economics of the Environment. London: Macmillan, 1971. A reprint of *The Swedish Journal of Economics* 73 (March 1971).

Borcherding, T. "Liability in Law and Economics: Note." *American Economic Review* 60 (December 1970): 946-48.

Bowen, H.R. *Toward Social Economy.* New York: Rinehart, 1948.

Bramhall, D.F. and Mills, E.S. "A Note on the Symmetry between Fees and Payments." *Water Resources Research* 2 (Third Quarter, 1966): 615-16.

Brownlee, O.H. "The Economics of Government Expenditures: Using Market Mechanisms in Making Government Expenditure Decisions." *American Economic Review* 49 (March-May 1959): 359-67.

Buchanan, J.M. *The Demand and Supply of Public Goods.* Chicago: Rand McNally, 1968.

_____. "An Economic Theory of Clubs." *Economica* 32 (February 1965): 1-13.

_____. "External Diseconomies, Corrective Taxes and Market Structure." *American Economic Review* 59 (March 1969): 174-77.

_____. "Joint Supply, Externalities, and Optimality." *Economica* 33 (November 1966): 404-15.

_____. "Politics, Policy and the Pigouvian Margins." *Economica* 29 (February 1962): 17-28.

_____, and Kafoglis, M.Z. "A Note on Public Goods Supply." *American Economic Review* 53 (June 1963): 403-14.

_____, and Stubblebine, W.C. "Externality." *Economica* 29 (November 1962): 371-84.

Burkhead, J. and Miner, J. *Public Expenditure.* Chicago: Aldine, 1971.

Burmeister, E. and Dobell, A.R. "Guidance and Optimal Control of Free-Market Economies: A New Interpretation." *IEEE Transactions on Systems, Man, and Cybernetics,* vol. SMC-2 (January 1972): 9-15.

Burrows, P. "On External Costs and the Visible Arm of the Law." *Oxford Economic Papers* 22 (March 1970): 39-55.

Bush, P.D. "The Normative Implications of Positive Analysis." Western Economic Association Meetings, August 1968, (mimeo).

Calabresi, G. "The Decision for Accidents: An Approach to Nonfault Allocation of Costs." *Harvard Law Review* 78 (1965): 713-30.

———. "Transactions Costs, Resource Allocation, and Liability Rules—A Comment." *Journal of Law and Economics* (April 1968): 67-73.

California Air Resources Board. Annual Reports. *Air Pollution in California.* Sacramento.

Center for the Study of Responsive Law. A Ralph Nader Task Force Report. *Water Wasteland.* New York: Bantam Books, 1971.

Cicchetti, C.J. and Freeman, A.M., III. "Option Demand and Consumer Surplus." *Quarterly Journal of Economics* 86 (August 1971): 528-39.

Coase, R.H. "The Problem of Social Cost." *Journal of Law and Economics* 3 (October 1960): 1-44.

Commoner, B. *The Closing Circle, Nature, Man and Technology.* New York: Alfred A. Knopf, 1971.

Comptroller General of the United States. *Examination into the Effectiveness of the Construction Program for Abating, Controlling and Preventing Water Pollution.* Washington, D.C.: United States Government Printing Office, 1969.

Converse, A.O. "On the Extension of Input-Output Analysis to Account for Environmental Externalities." *American Economic Review* 61 (1971): 197-98.

———. "Optimum Number and Location of Treatment Plants." *Journal of the Water Pollution Control Federation* 44 (August 1972): 1629-36.

Council on Environmental Quality. Annual Reports. *Environmental Quality.* Washington, D.C.: United States Government Printing Office, beginning 1970.

Crocker, T.D. "The Structuring of Atmospheric Pollution Control Systems." *The Economics of Air Pollution.* Edited by H. Wolozin. New York: W.W. Norton, 1966.

Crowe, B.L. "The Tragedy of the Common Revisited." *Science* 166 (November 28, 1969): 1103-07.

Dahmen, E. "Environmental Control and Economic Systems." *The Swedish Journal of Economics* 75 (March 1971): 67-75.

Dales, J.H. "Land, Water, and Ownership." *Canadian Journal of Economics* 34 (November 1968).

———. *Pollution, Property and Prices, An Essay in Policymaking and Economics.* Toronto: University of Toronto Press, 1968.

D'Arge, R.C. "Essay on Economic Growth and Environmental Quality." *Swedish Journal of Economics* 73 (March 1971): 25-41.

———, and Hunt, E.K. "Environmental Pollution, Externalities and Conventional Economic Wisdom: A Critique." *Environmental Affairs* 1 (June 1971).

Davidson, B. and Bradshaw, R.W. "Thermal Pollution of Water Systems." *Environmental Science and Technology* 1 (August 1967): 618-30.

Davidson, P.; Adams, F.G.; and Seneca, J. "The Social Value of Water Recreational Facilities Resulting from an Improvement in Water Quality: The Delaware Estuary." Kneese, A.V. and Smith, S.C. *Water Research*. Baltimore: Johns Hopkins Press, 1966.

Davis, R.K. *The Range of Choice in Water Management*. Baltimore: Johns Hopkins Press for Resources for the Future, 1968.

Davis, O.A. and Kamien, M.I. "Externalities, Information and Alternative Collective Action." *The Analysis of Public Expenditures: The P.P.B. System*. Washington, D.C.: United States Government Printing Office, 1969.

_____ and Whinston, A. "Externalities, Welfare and the Theory of Games." *Journal of Political Economy* 70 (June 1962): 241-62.

_____ and Whinston, A. "On the Distinction between Public and Private Goods." *American Economic Review* 57 (May 1967): 360-73.

_____ and Whinston, A. "On Externalities, Information and the Government-Assisted Invisible Hand." *Economica* 33 (August 1966): 303-18.

_____ and Whinston, A. "Piecemeal Policy on the Theory of Second Best." *Review of Economic Studies* 34 (July 1967): 323-31.

_____ and Whinston, A. "Welfare Economics and the Theory of Second Best." *Review of Economic Studies* 32 (January 1965): 1-14.

Decker, R.S. "The Dynamics of Unconventional Motor Vehicles and the Reduction of Air Pollution." *Western Economic Journal* 8 (December 1970): 357-63.

Demsetz, H. "The Exchange and Enforcement of Property Rights." *Journal of Law and Economics* 7 (October 1964): 11-26.

_____. "Theoretical Efficiency in Pollution Control: Comments on Comments." *Western Economic Journal* 9 (December 1971): 444-46.

_____. "Toward a Theory of Property Rights." *American Economic Review* 57 (1967).

Diamond, P.A. "Consumption Externalities and Imperfect Corrective Pricing." *Bell Journal of Economics and Management Science* 4 (Autumn 1973): 526-38.

_____ and Mirrlees, J. "Aggregate Production with Consumption Externalities." *Quarterly Journal of Economics* 87 (February 1973): 1-24.

Dolan, E.G. *TANSTAAFL, The Economic Strategy for Environmental Crisis*. New York: Holt, Rinehart and Winston, Inc., 1971.

Dolbear, F.T. "On the Theory of Optimal Externality." *The American Economic Review* 57 (March 1967): 90-103.

Dorfman, R. "Conceptual Model of a Regional Water Quality Authority." *Models for Managing Regional Water Quality*. Edited by R. Dorfman, H. Jacoby, and H.A. Thomas. Cambridge, Massachusetts, 1973.

_____. "Discussion." *American Economic Review* 63 (May 1973): 253-56.

Eckstein, O. "A Survey of the Theory of Public Expenditures Criteria." National Bureau of Economic Research. *Public Finances: Needs, Sources and Utilization*. Princeton, New Jersey: Princeton University Press, 1961.

Ellis, H. and Fellner, W. "External Economies and Diseconomies." *American Economic Review* 23 (September 1943): 493-511. Reprinted in *Readings in Price Theory*. Chicago: Irwin, 1952.

Environmental Protection Agency, Water Quality Office. *Cost of Clean Water*. Washington, D.C.: United States Government Printing Office, 1971.

_____. *Stochastic Modeling for Water Quality Management*. Washington, D.C.: United States Government Printing Office, 1971.

Evans, A.W. "Private Good, Externality, Public Good." *Scottish Journal of Political Economy* 17 (February 1970): 79-89.

Fabricant, N. and Hallman, R.M. *Toward a Rational Power Policy, Energy, Politics and Pollution, a Report of the Environmental Protection Agency of the City of New York*. New York: George Braziller, 1971.

Federal Water Pollution Control Administration. *Delaware Estuary Comprehensive Study: Preliminary Report and Findings*. Philadelphia, 1966.

Feldstein, M. "Distributional Equity and the Optimal Structure of Public Prices." *American Economic Review* 62 (March 1972): 32-36.

Forte, F. and Buchanan, J.M. "The Evaluation of Public Services." *Journal of Political Economy* 69 (April 1961): 107-21.

Freeman, A.M., III. "Bribes and Charges: Some Comments." *Water Resources Research* 3 (Fall Quarter 1967).

_____; Haveman, R.H.; and Kneese, A.V. *The Economics of Environmental Policy*. New York: Wiley, 1973.

Furubotn, E. and Pejovich, S. "Property Rights and Economic Theory: A Survey of Recent Literature." *Journal of Economic Literature* 10 (December 1972): 1143.

Gerhardt, P. "Incentives to Air Pollution Control." *Law and Contemporary Problems* 33 (1968): 358-68.

Gifford, A., Jr., and Stone, C.C. "Externalities, Liability and the Coase Theorem: A Mathematical Analysis." *Western Economic Journal* 11 (September 1973): 260-69.

Goetz, C.J. and Buchanan, J.M. "External Diseconomies in Competitive Supply." *American Economic Review* 61 (December 1971): 833-90.

Goldman, M.I. "The Convergence of Environmental Disruption." *Science* 170 (October 2, 1970): 37-42.

_____. "Growth and Environmental Problems of Noncapitalists Nations." *Challenge* 16 (July-August 1973): 45-51.

Graff, J. de van. *Theoretical Welfare Economics*. Cambridge: Cambridge University Press, 1957.

Graves, G.W.; Hatfield, G.B.; and Whinston, A. "Water Pollution Control Using By-Pass Piping." *Water Resources Research* 5 (February 1969): 13-47.

Green, H.A.J. "The Social Optimum in the Presence of Monopoly and Taxation." *Review of Economic Studies* 29 (1962): 66-78.

Hadley, G. *Nonlinear and Dynamic Programming*. Reading, Massachusetts: Addison-Wesley, 1964.

Haefele, E.T. "A Utility Theory of Representative Government." *American Economic Review* 61 (June 1971): 350-67.

Hahn, F.H. "Equilibrium Dynamics with Heterogeneous Capital Goods." *Quarterly Journal of Economics* 80 (1966): 633-46.

Hardin, G. "The Tragedy of the Commons." *Science* 162 (December 13, 1968): 1243-48.

Harris, C.M., ed. *Handbook of Noise Control.* New York: McGraw-Hill, 1957.

Haveman, R.H. and Margolis, J., eds. *Public Expenditures and Policy Analysis.* Chicago: Markham Publishing Company, 1970.

Head, J.G. "Public Goods and Public Policy." *Public Finance* 17 (1962): 197-221.

Helfrich, H.W., Jr., ed. *Agenda for Survival: The Environmental Crisis.* New Haven: Yale University Press, 1971.

Herfindahl, O.C. and Kneese, A.V. *Quality of the Environment: An Economic Approach to Some Problems in Using Land, Water and Air.* Baltimore: Johns Hopkins Press for Resources for the Future, 1965.

Hicks, J.R. "The Foundations of Welfare Economics." *Economic Journal* 49 (December 1939): 696-712.

_____. "The Four Consumer's Surpluses." *Review of Economic Studies* 11 (1943): 31-41.

Hoch, I. "Economic Analysis of Wilderness Areas." *Wilderness and Recreation— A Report on Resources; Values and Problems, ORRRC Study Report No. 3.* Washington, D.C., 1962.

Hunt, E.K. and D'Arge, R.C. "On Lemmings and Other Acquisitive Animals." *Journal of Economic Issues* 7 (June 1973): 337-53.

Hurwicz, L. "Optimality and Informational Efficiency in Resource Allocation Processes." *Mathematical Methods in the Social Sciences, 1959: Proceedings of the First Stanford Symposium.* Edited by K. Arrow and S. Karlin. Stanford: Stanford University Press, 1960.

Institute of Public Administration. *Governmental Approaches to Air Pollution Control.* Washington, D.C.: United States Government Printing Office, 1971.

Intriligator, M. *Mathematical Optimization and Economic Theory.* Englewood Cliffs, New Jersey: Prentice-Hall, 1971.

Jackson, W.E. and Wohlers, H.C. "Regional Air Pollution Control Costs." *Journal of the Air Pollution Control Association* 22 (September 1972): 679-84.

Jacoby, H. and Loucks, D.P. "The Combined Use of Optimization and Simulation Models in River Basin Planning." *Water Resources Research* 8 (December 1972).

Jarrett, H., ed. *Environmental Quality.* Baltimore: Johns Hopkins Press for Resources for the Future, 1966.

Kahn, R.F. "Some Notes on Ideal Output." *Economic Journal* 45 (March 1935): 1-35.

Kamien, M.J.; Schwartz, N.L.; and Dolbear, F.T., Jr. "Asymmetry between Bribes and Charges." *Water Resources Research* 2 (First Quarter, 1966): 147-57.

Karlin, S. *Mathematical Methods and Theory in Games, Programming and Economics*, Vol. I. Reading, Massachusetts: Addison-Wesley, 1959.

Keeler, E.; Spence, M.; and Zeckhauser, R. "The Optimal Control of Pollution." *Journal of Economic Theory* 4 (1971): 19-34.

Kneese, A.V. "Air Pollution—General Background and Some Economic Aspects." *The Economics of Air Pollution.* Edited by H. Wolozin. New York: W.W. Norton, 1966.

————. *Approaches to Regional Water Quality Management.* Baltimore: Johns Hopkins Press for Resources for the Future, 1967.

————. *The Economics of Regional Water Quality Management.* Baltimore: Johns Hopkins Press for Resources for the Future, 1964.

————. "Environmental Pollution: Economics and Policy." *American Economic Review* 61 (May 1971): 153-66.

————. *Water Pollution: Economic Aspects and Research Needs.* Baltimore: Johns Hopkins Press for Resources for the Future, 1962.

————; Ayres, R.U.; and D'Arge, R.C. *Economics and the Environment: A Materials Balance Approach.* Baltimore: Johns Hopkins Press for Resources for the Future, 1970.

———— and Bower, B.T., eds. *Environmental Quality Analysis.* Baltimore: Johns Hopkins Press for Resources for the Future, 1972.

———— and Bower, B.T. *Managing Water Quality: Economics, Technology, Institutions.* Baltimore: Johns Hopkins Press for Resources for the Future, 1968.

———— and D'Arge, R.C. "Pervasive External Costs and the Response of Society." 91st U.S. Congress, Joint Economic Committee, Subcommittee on Economy in Government. *The Analysis and Evaluation of Public Expenditures: The P.P.B. System.* Washington, D.C.: United States Government Printing Office, 1969.

Knight, F.H. "Some Fallacies in the Interpretation of Social Cost." *Quarterly Journal of Economics* 37 (August 1924): 582-606.

Kohn, R.E. "Input-Output Analysis and Air Pollution Control." Paper presented at the National Bureau of Economic Research-Resources for the Future Conference on The Economics of the Environment, Chicago, 1970.

Koopmans, T.C. "Objectives, Constraints, and Outcomes in Optimal Growth Models." *Econometrica* 35 (January 1967): 1-15.

Krutilla, J.V. "Conservation Reconsidered." *The American Economic Review* 57 (September 1967): 777-86.

————. "Some Environmental Effects of Economic Development." *Daedalus* (Fall 1970): 1058-70.

Krutilla, J.V. "Welfare Aspects of Benefit-Cost Analysis." *Journal of Political Economy,*69 (June 1961): 226-35.

Kuhn, H. and Tucker, A. "Non-linear Programming." *Proceedings of the Second Berkeley Symposium on Mathematical Statistics and Probability.* Edited by J. Neyman. Berkeley: University of California Press, 1951.

Landsberg, H.H. "The U.S. Resource Outlook, Quantity and Quality." *Daedalus* (Fall 1967): 1034-57.

Lange, O. "The Foundations of Welfare Economics." *Econometrica* 10 (July-October 1942): 215-28.

Lave, L. and Seskin, E. "Air Pollution and Human Health." *Science* 21 (August 1970): 723-33.

_____ and Seskin, E. "Health and Air Pollution." *The Swedish Journal of Economics* 73 (March 1971): 76-95.

Leontief, W. "Environmental Repercussions and the Economic Structure: An Input-Output Approach." *Review of Economics and Statistics* 52 (August 1970): 374-81.

Lipsey, R.G. and Lancaster, K. "The General Theory of Second Best." *Review of Economic Studies* 24 (February 1956): 11-32.

Little, I.M.D. *A Critique of Welfare Economics.* Oxford: Oxford University Press, 1957.

Loucks, D.P. and Jacoby, H. "Flow Regulation for Water Quality." *Models for Managing Regional Water Quality.* Edited by R. Dorfman, H. Jacoby, and H.A. Thomas. Cambridge, Massachusetts, 1973.

Maas, A. "Benefit Cost Analysis—Its Relevance for Public Investment Decisions." *The Quarterly Journal of Economics* 80 (May 1966): 208-26.

McDaniel, P.R. and Kaplinsky, A.S. "The Use of the Federal Income Tax System to Combat Air and Water Pollution: A Case Study in Tax Expenditures." *Boston College Industrial and Commercial Law Review* 12 (February 1971).

McGuire, M.C. and Aaron, H. "Efficiency and Equity in the Optimal Supply of a Public Good." *Review of Economics and Statistics* 51 (February 1969): 31-39.

McKean, R.N. *Efficiency in Government Through Systems Analysis, with Emphasis on Water Resources Development.* New York: John Wiley and Sons, 1958.

_____. "The Unseen Hand in Government." *The American Economic Review* 55 (June 1965): 496-507.

McManus, M. "Private and Social Costs in the Theory of Second Best." *Review of Economic Studies* 34 (July 1967): 317-21.

Marchand, J.R. and Russell, K.P. "Externalities, Liability, Separability, and Resource Allocation." *American Economic Review* 63 (September 1973): 611-20.

Margolis, J. "A Comment on the Pure Theory of Public Expenditure." *Review of Economics and Statistics* 37 (November 1955): 347-49.

_____ and Vincent, P.E. *External Economic Effects–An Analysis and Survey*. Stanford: Engineering Economic Systems, 1966.

Marshall, A. *Principles of Economics*, 8th ed. London: Macmillan, 1925.

Meade, J.E. "External Economies and Diseconomies in a Competitive Situation." *The Economic Journal* 62 (1952): 54-67.

Mills, E.S. "Economic Incentives in Air-Pollution Control." *The Economics of Air Pollution*. Edited by H. Wolozin. New York: W.W. Norton, 1966.

Mishan, E.J. *The Costs of Economic Growth*. New York: Frederick Praeger, 1967.

_____ . "Pangloss on Pollution." *Swedish Journal of Economics* 73 (March 1971): 111-20. Reprinted in Bohm, P. and Kneese, A.V., *The Economics of Environment*. London: Macmillan, 1971.

_____ . "Pareto Optimality and the Law." *Oxford Economic Papers* 19 (November 1967): 255-87.

_____ . "The Postwar Literature on Externalities: An Interpretive Essay." *Journal of Economic Literature* 9 (March 1971): 1-28.

_____ . "Reflections on Recent Developments in the Concept of External Effects." *Canadian Journal of Economics* 31 (February 1965): 3-34.

_____ . "The Relationship between Joint Products, Collective Goods and External Effects." *Journal of Political Economy* 77 (May 1969): 329-48.

_____ . "Second Thoughts on Second Best." *Oxford Economic Papers*, 14 (October 1962): 205-17.

_____ . *Technology and Growth: The Price We Pay*. New York: Frederick A. Praeger, 1970.

_____ . "Welfare Criteria for External Effects." *American Economic Review* 51 (September 1961): 594-613.

_____ . "What is Producers' Surplus?" *American Economic Review* 58 (December 1968): 1269-82.

Morrison, C.C. "Generalizations on the Methodology of the Second Best." *Western Economic Journal* 6 (1968): 112-20.

Mumey, G.A. "The 'Coase Theorem': A Reexamination." *Quarterly Journal of Economics* 85 (November 1971): 718-23.

Murakami, Y. and Negishi, T. "A Note on a Formulation of External Economy." *International Economic Review* 5 (September 1964): 328-34.

Musgrave, R.A. "Cost-Benefit Analysis and the Theory of Public Finance." *Journal of Economic Literature* 7 (September 1969): 797-806.

_____ . *The Theory of Public Finance*. New York: McGraw-Hill, 1959.

National Academy of Sciences, the Committee on Pollution. Report to the Federal Council for Science and Technology. *Waste Management and Control*. Washington, D.C.: National Research Council, 1966.

Negishi, T. "Welfare Economics and Existence of an Equilibrium for a Competitive Economy." *Metroeconomica* 11 (1960): 92-97.

Noll, R.G. and Trijonis, J. "Mass Balance, G.E., and Environmental Externalities." *American Economic Review* 61 (1971): 730-35.

Nutter, G.W. "The Coase Theorem on Social Cost: A Footnote." *Journal of Law and Economics* 11 (October 1968): 503-07.

Oakland, W.H. "Joint Goods." *Economica* 36 (August 1969), pp. 253-68.

Olson, M., Jr., *The Logic of Collective Action.* Cambridge, Massachusetts: Harvard University Press, 1965.

_____ and Zeckhauser, R. "The Efficient Production of External Economies." *American Economic Review* 60 (June 1970): 512-17.

Pauly, M.V. "Clubs, Commonality, and the Core: An Integration of Game Theory and the Theory of Public Goods." *Economica* 34 (August 1967): 314-24.

Pearse, P.H. "A New Approach to the Evaluation of Non-Priced Recreation Resources." *Land Economics* 44 (February 1968): 87-99.

Pitchford, J. "Population and Optimal Growth." *Econometrica* 40 (January 1972): 109-36.

Pigou, A.C. *The Economics of Welfare*, 2d ed. London: Macmillan, 1924.

Plott, C.R. "Externalities and Corrective Taxes." *Economica* 33 (February 1966): 84-87.

Plourde, C.G. "A Simple Model of Replenishable Natural Resource Exploitation." *American Economic Review* 60 (June 1970): 518-22.

Pollack, W.L. "Legal Boundaries of Air Pollution Control—State and Local Legislative Purpose and Techniques." *Law and Contemporary Problems* 33 (1968): 331-57.

Pontryagin, L.S.; Boltyanskii, V.G.; Gamkrelidze, R.V.; and Meshchenko, E.F. *The Mathematical Theory of Optimal Processes.* New York: Interscience Publishers. 1962.

President's Council on Environmental Quality. *Environment Quality.* Annual Reports. Washington, D.C.: United States Government Printing Office, Beginning 1970.

President's Science Advisory Committee, Environmental Panel. J.W. Tukey, Chairman. *Restoring the Quality of the Environment.* Washington, D.C.: White House, 1965.

Prest, A.R. and Turvey, R. "Cost-Benefit Analysis: A Survey." *The Economic Journal* 74 (December 1965): 683-735.

Quirk, J. and Saposnik, R. *Introduction to General Equilibrium Theory and Welfare Economics.* New York: McGraw-Hill, 1968.

Ramsey, F.P. "A Mathematical Theory of Saving." *Economic Journal* (December 1928): 543-59.

Reder, M.W. *Studies in the Theory of Welfare Economics.* New York: Columbia University Press, 1947.

Reed, K.R. "Economic Incentives for Pollution Abatement: Applying Theory to Practice." *Arizona Law Review* 12 (1970): 517-18.

Reich, C.A. "The New Property." *Yale Law Journal* 73 (April 1964): 733-87.

Reitze, A.W. and Reitze, G. "Tax Incentives Don't Stop Pollution." *American Bar Association Journal* 57 (February 1971).

Renshaw, E. "A Note on the Measurements of the Benefits from Public Investment in Navigation Projects." *American Economic Review* 47 (June-December 1957): 652-62.

Ridker, R.G. *Economic Costs of Air Pollution*, New York: Frederic Praeger, 1967.

_____ and Henning, J.A. "The Determinants of Residential Property Values with Special Reference to Air Pollution." *The Review of Economics and Statistics* 49 (1967): 246-57.

Roberts, M.J. "Organizing Water Pollution Control: The Scope and Structure of River Basin Authorities." *Public Policy* 19 (Winter 1971): 79-141.

_____ . "River Basin Authorities: A National Solution to Water Pollution." *Harvard Law Review* 83 (1970): 1527-56.

Rothenberg, J. "The Economics of Congestion and Pollution: An Integrated View." *American Economic Review* 60 (May 1970): 114-21.

Ruff, L. "The Economic Common Sense of Pollution." *The Public Interest* (Spring 1970): 69-85.

Russell, C.S. "Application of Microeconomic Models to Regional Environmental Quality Management." *American Economic Review* 63 (May 1973): 236-43.

_____ . "Models for Investigation of Industrial Response to Residuals Management Actions." *The Economics of Environment.* Edited by P. Bohm and A.V. Kneese. London: Macmillan, 1971.

_____ and Spofford, W.O., Jr. "A Qualitative Framework for Residuals Management Decisions." *Environmental Quality Analysis: Research Studies in the Social Sciences.* Edited by A.V. Kneese and B.T. Bower. Baltimore: Johns Hopkins Press for Resources for the Future, 1971.

Samuelson, P.A. "Aspects of Public Expenditure Theories." *Review of Economics and Statistics* 40 (November 1958): 332-38.

_____ . "Contrast between Welfare Conditions for Joint Supply and for Public Goods." *Review of Economics and Statistics* 51 (February 1969): 26-30.

_____ . "Diagrammatic Exposition of a Theory of Public Expenditure." *Review of Economics and Statistics* 37 (November 1955): 350-56.

_____ . "Further Comment on Welfare Economics." *American Economic Review* 33 (September 1943): 604-07.

_____ . "The Pure Theory of Public Expenditure." *The Review of Economics and Statistics* 36 (November 1954): 387-89.

_____ and Solow, R.M. "A Complete Model Involving Heterogeneous Capital Goods." *Quarterly Journal of Economics* 70 (1956): 537-62.

Sax, J.L. *Defending the Environment.* New York: Knopf, 1970.

Schall, L.D. "Technological Externalities and Resource Allocation." *Journal of Political Economy* 79 (September/October 1971): 983-1001.

Schelling, T.C. "On the Ecology of Micromotives." *The Public Interest* 25 (Fall 1971): 59-98.

Scitovsky, T. "External Diseconomies in the Modern Economy." *The Western Economic Journal* 4 (Summer 1966): 197-202.

Scitovsky, T. "Two Concepts of External Economies." *Journal of Political Economy* 62 (April 1954): 143-51.

Shell, K., ed. *Essays on the Theory of Optimal Economic Growth*. Cambridge, Massachusetts: M.I.T. Press, 1967.

Shoup, D.C. "Theoretical Efficiency in Pollution Control: Comment." *Western Economic Journal* 9 (September 1971): 310-13.

Smith, V.L. "Dynamics of Waste Accumulation: Disposal versus Recycling." *Quarterly Journal of Economics* 86 (1972): 600-16.

_____. "Economics of Production from Natural Resources." *American Economic Review* 58 (June 1968): 409-31.

Smith, E.T. and Morris, A.R. "Systems Analysis for Optimal Water Quality Management." *Water Pollution Control Journal* 41 (September 1969): 1635-43.

Solow, R. "The Economist's Approach to Pollution and Its Control." *Science* 173 (1971): 498-503.

Starrett, D. "Fundamental Nonconvexities in the Theory of Externalities." *Journal of Economic Theory* 4 (April 1972): 180-99.

Steiner, P.O. "Choosing Among Alternative Public Investments in the Water Resource Field." *American Economic Review* 49 (June-December 1959): 893-916.

Stern, A., ed. *Air Pollution and Its Effects*, Vols. I, II, III. New York: Academic Press, 1968.

Stigler, G.J. "The New Welfare Economics." *American Economic Review* 33 (June 1943): 355-59.

Streeter, H.W. and Phelps, E.B. *A Study of the Pollution and Natural Purification of the Ohio River—III: Factors Concerned in the Phenomena of Oxydation and Recreation*. Public Health Bulletin No. 146. Washington, D.C.: United States Government Printing Office, February 1925.

The Study of Critical Environmental Problems Sponsored by the Massachusetts Institute of Technology. Report. *Man's Impact on the Global Environment*. Cambridge, Massachusetts: M.I.T. Press, 1970.

Surrey, S.S. "Tax Incentives as a Device for Implementing Government Policy: A Comparison with Direct Government Expenditures." *Harvard Law Review* 83 (February 1970): 706-13.

Teller, A.A. "Air Pollution Abatement: Economic Rationality and Reality." *Daedalus* (Fall 1967): 1082-98.

_____. "The Use of Linear Programming to Estimate the Cost of Some Alternative Air Pollution Abatement Policies." *Proceedings, IBM Scientific Computing Symposium on Water and Air Resource Management*. Yorktown Heights, New York; 1967.

Thomas, H.A., Jr. "The Animal Farm: A Mathematical Model for the Discussion of Social Standards for the Control of the Environment." *Quarterly Journal of Economics* (February 1963). Reprinted in *Economics of the Environment:*

Selected Readings. Edited by R. Dorfman and N.S. Dorfman. New York: W.W. Norton, 1972.

Tolley, G.S. "McKean on Government Efficiency." *Review of Economics and Statistics* 41 (November 1959): 446-48.

TRW, Incorporated. "Air Quality Implementation Planning Program," for Environmental Protection Agency, 1. November, 1970.

Tulloch, G. "Problems in the Theory of Public Choice, Social Cost and Government Action." *The American Economic Review* 59 (May 1969): 189-97.

Turvey, R. "On Divergences between Social Cost and Private Cost." *Economica* 30 (August 1963): 309-13.

Vincent, P.E. "Reciprocal Externalities and Optimal Input and Output Levels." *American Economic Review* 59 (1969): 976-83.

Viner, J. "Cost Curves and Supply Curves." *Zeitschrift für Nationalökonomie* 3 (September 1931): 23-46.

Wellisz, S. "On External Diseconomies and the Government-Assisted Invisible Hand." *Economica* 31 (November 1964): 345-62.

White, L., Jr. "The Historical Roots of Our Ecological Crisis." *Science* 155 (March 10, 1967): 1203-07.

Williams, A. "The Optimal Provision of Public Goods in a System of Local Government." *Journal of Political Economy* 74 (January 1966).

Williamson, O.E. "Peak-Load Pricing and Optimal Capacity Under Indivisibility Constraints." *American Economic Review* 56 (September 1966).

Wolozin, H., ed. *The Economics of Air Pollution.* New York: W.W. Norton, 1966.

_____. "The Economics of Air Pollution: Central Problems." *Law and Contemporary Problems* 33 (1968): 227-38.

Worchester, D.A., Jr. "Pecuniary and Technological Externality, Factor Rents, and Social Costs." *American Economic Review* 59 (December 1969): 873-85.

Wright, C. "Some Aspects of the Use of Corrective Taxes for Controlling Air Pollution Emissions." *Natural Resources Journal* 9 (January 1969): 63-82.

Zerbe, R.O. "Theoretical Efficiency in Pollution Control." *Western Economic Journal* 8 (December 1970): 364-76.

_____. "Theoretical Efficiency in Pollution Control: Reply." *Western Economic Journal* 9 (September 1971): 314-17.

Index

abatement, 23, 56; interfirm efficient, 72; intrafirm efficient, 67ff, 72
allocative game, 6
appropriability, 22
Arrow, K.J., 53, 54, 86
Arrow's assumption two, 4, 10, 19
Ayres, R.U., 125

bargaining, 77ff, 109
Bator, F.M., 21, 22
Baumol, W.J., 49, 97, 114, 120, 123
Baumol, W.J., and Oates, W.E., 120
benefits function, 56
bilateral effects, 48
Buchanan, J.M., 22, 23, 94
Buchanan, J.M. and Kafoglis, M.Z., 48, 49
Buchanan, J.M. and Stubblebine, W.C., 20, 22, 23, 94, 103
Burrows, P., 81

Calabresi, G., 84
charges, 110
Coase, R., 78, 91, 113
Coase's theorem, 80; and liability rules, 83
corner solution, 8, 10, 29, 39, 100
cost benefit ratio, 105
Crocker, T.D., 72

Dales, J.H., 120, 123
damage, tangible and intangible, 67
damages function, 56
D'Arge, R.C., 125
Davis, O.A., 89, 93, 97, 106
demand for pollution, 62
Demsetz, H., 54, 87
Diamond, P.A., 96, 97, 106
Diamond, P.A. and Mirrlees, J., 49, 50, 81
directives, 100
divorce of scarcity from ownership, 21
Dolbear, F.T., 23, 81, 96
dynamic models, 130

efficiency, interfirm, 72; intrafirm, 72
Ellis, H. and Fellner, W., 21
equilibrium price, 16
equity, 116

exclusion, 22
external scale economies, 100
externalities, 20; pecuniary, 19; technological, 20, 26, 30, 42; inframarginal, 20; marginal, 20; ownership, 21

free rider problem, 40, 41, 88
Freeman, A.M., III, 116

Genossenschaften, 121
Goetz, C.J. and Buchanan, J.M., 98
Goldman, M.I., 1

Hahn, F., 132
Hamiltonian function, 132; interior solutions, 133; boundary solutions, 135
Head, J.G., 21

independence, assumption of, *see* Arrow's assumption two
input requirement functions, 9, 12, 15, 16, 37, 41
interdependence, 20, 21, 22
inversion, 64
involuntary transfers, 91
isopollution-reduction curves, 67

joint supply, 22, 23

Kamien, M.J., Schwartz, N.L. and Dolbear, F.T., 116
Kneese, A.V., 121, 125
Kneese, A.V. and Bower, B.T., 119, 121, 123
Kuenne, R., 13
Kuhn-Tucker theorem, 6, 10, 34, 39

Langham, M.L., Headley, J.C. and Edwards, W.F., 117
Legrangian function, 6, 10, 13, 34, 35, 68, *see also* payoff function
Legrangian multiplier, 10, 13, 38
Leontief, W., 125
liability assignment, 77ff
lighthouse example, 86

Marchand, J.R., 84, 87

175

About the Author

Daniel T. Dick is Assistant Professor of Economics at the University of Santa Clara. He received the Ph.D. in 1970 from Claremont Graduate School, where he was an NDEA and Haynes Fellow. He is a member of the American Economic Association and the American Association of University Professors; his research involves applied microeconomics. Dr. Dick has been Visiting Assistant Professor at the California Institute of Technology and an independent consultant for the State of California Department of Transportation.